Microwave Know How

for the Radio Amateur

Edited by Andy Barter, G8ATD

Radio Society of Great Britain

Published by The Radio Society of Great Britain, 3 Abbey Court, Fraser Road, Priory Business Park, Bedford, MK44 3WH.

First published 2010

ISBN: 9781 9050 8656 3

Publisher's note

The opinions expressed in this book are those of the authors and not necessarily those of the RSGB. While the information presented is believed to be correct, the authors, the publisher and their agents cannot accept responsibility for the consequences arising for any inaccuracies or omissions.

Cover design: Kim Meyern

Production: Mark Allgar, M1MPA

Typography: Andy Barter, K M Publications, Luton

Cover photograph kindly supplied by VHF Communications Magazine

Printed in Great Britain by Page Bros of Norwich

Acknowledgements

The articles in this book have been supplied by radio amateurs from around the world. The editor would like to thank the following people and organisations for their contributions:

Ralph Berres, DF6WU

Kent Britain, 2E0VAA/WA5VJB

JOhn Fielding, ZS5JF

José Geraldo Chiquito

Dom Dehays, F6DRO

Stephen Hayman, ZL1TPH

Sam Jewell, G4DDK

Ed Johnson, AD5MQ

Michael Kohla, DL1YMK

Gunthard Kraus, DG8GB

Alexander Meier, DG6RBP

Roger Ray, G8CUB

Franco Rota, I2FHW

Matjaz Vidmar, S53MV

Paul Wade, W1GHZ

Henning C Weddig, DK5LV

Dr John Worsnop, G4BAO

RadCom

UKW Berichte Magazine

VHF Communications Magazine

Scatterpoint, the magazine of the UK Microwave Group

Contents

Preface

It has been five years since I edited Microwave Project 2. Since then the devices and equipment available to amateurs interested in the microwave bands has continued to increase. This book contains some of the things that are being used right now and some the surplus equipment that has come onto the market that can be modified for use on the microwave bands.

The articles on surplus equipment have been taken from Scatterpoint, the magazine of the UK Microwave Group, this is a very active group an I recommend that any amateurs in the UK should get along to one of the meeting that are held around the country to see how useful being part of this group can be to get the most out of the microwave bands. There are similar groups in many other countries that I an sure would be just as helpful if you do not live in the UK.

So that readers will not think this is just a face-lifted version of the earlier Microwave Projects books it has been decided to give this book a new more descriptive title.

Many of the articles require a PCB, the artwork in this book is not reproduced to exact size, so if you are going to make you own PCBs please take care to scale the artwork correctly. I recommend that you take an easier option because most of the authors can supply PCBs and I have included contact details where possible. Many of the contacts are email addresses or web sites, unfortunately these have a habit of changing, they were current at the time of editing (January 2010). If you have difficulties with PCBs, finding components or contact details please contact me and I will try to help.

Andy Barter, G8ATD, Email: andy@vhfcomm.co.uk

Chapter 1

Antennas

In this chapter :

- Weatherproof microwave antennas
- 6cm antenna
- Vivaldi antennas

Antennas are one of the features of a microwave station that differentiate it from other frequencies. We are lucky that incredible gain can be achieved and they can be placed at many wavelengths above the ground with ease. Something that makes amateurs operating on other bands very envious.

Weatherproof antennas for UHF and lower microwave frequencies [1]

Radio amateurs, weather effects and antenna design

Quite frequently, radio amateur antenna design is limited to a few popular antenna types. Directional antenna designs are usually limited to Yagi antennas for the lower frequencies and parabolic dishes for the microwave frequency bands. The operation of a Yagi antenna is based on a collimating lens made of artificial dielectric like rods, loops, disks or helices. The basic design goal of all these slow-wave structures is to achieve the maximum antenna directivity with the minimum amount of material (metal).

The situation is actually made worse with the availability of inexpensive antenna simulation tools for home computers. The latter provide designs with fantastic gain figures using little hardware. Unfortunately, these results are barely useful in practice. Besides impedance matching problems, such designs are extremely sensitive to manufacturing tolerances and environmental conditions: reflections from nearby objects and accumulation of dirt or raindrops on the antenna structure.

The operation of a 2m (144MHz) Yagi with thin rods (or loops or helix) antenna will be completely disrupted if snow or ice accumulates on the antenna structure. Raindrops accumulated on the antenna structure will completely compromise the operation of a 23cm (1.3GHz) or 13cm (2.3GHz) Yagi antenna. Manufacturing tolerances limit practical Yagi antennas to frequencies below about 5GHz.

Professional VHF Yagi antennas use very thick rods to limit the effects of snow and ice on the antenna performance. All professional Yagi antennas above 300MHz are completely enclosed inside radomes made of insulating material that is transparent to radio waves. Since any radome includes some dielectric and the latter has a considerable effect on any Yagi antenna, a Yagi antenna has to be completely redesigned for operation inside a weatherproof enclosure. If some natural dielectric (radome) is added, then some artificial dielectric (Yagi rods) has to be removed to maintain the same focal length of the dielectric collimating lens.

Simply speaking, any serious design of a weatherproof Yagi antenna for frequencies above 300MHz is out-of-reach for most amateurs. Fortunately there exist other antenna solutions for

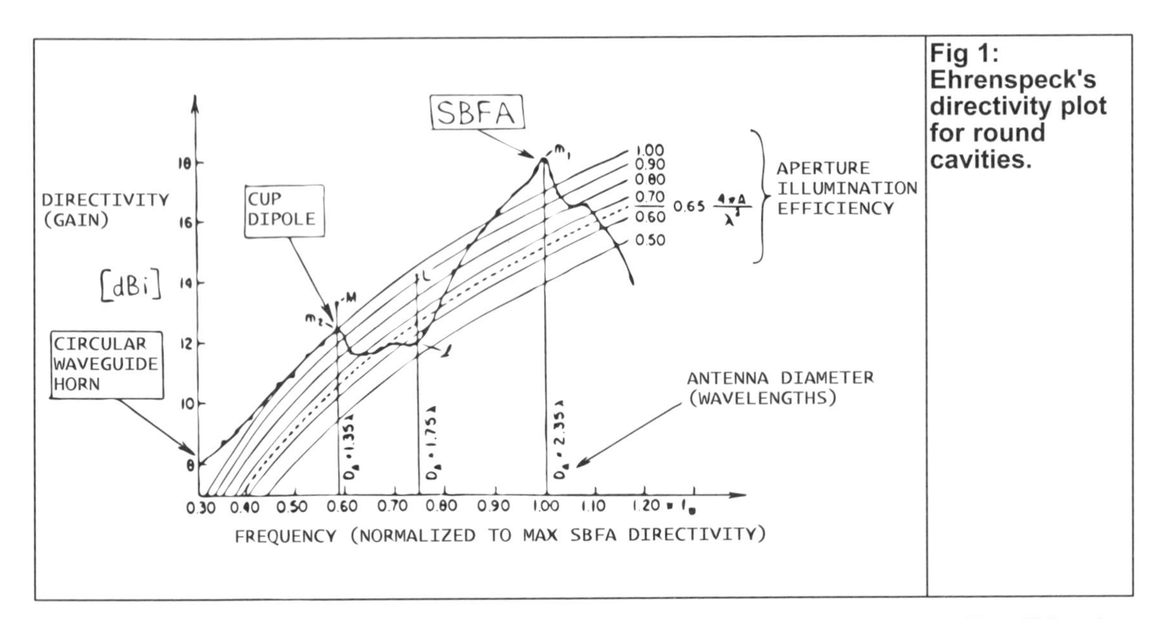

Fig 1: Ehrenspeck's directivity plot for round cavities.

amateur radio equipment that has to operate unattended on mountain tops like FM voice repeaters, ATV repeaters, packet radio nodes and microwave beacons. The most popular solutions are arrays of dipoles, quads, eights etc. The installation of the latter inside suitable weatherproofing radomes is much less critical than the weatherproofing of Yagis or helices.

Practical cavity antennas

Of course, a different antenna design approach may provide much better results, like considering the weatherproofing issue right from the beginning! Cavity antennas may initially require more metal for the same decibels of antenna gain. On the other hand, a cavity antenna is relatively easy to weatherproof: most of the radome is the metal cavity itself and just a relatively small radiating aperture has to be additionally protected with some transparent material.

Before deciding for a particular cavity antenna design, it makes sense to check well-known solutions. A comprehensive description of many different cavity antennas is given in the book [2]. A useful selection tool is Fig 1, the directivity plot as a function of cavity diameter as published by Ehrenspeck [3]:

Although Ehrenspeck's diagram is a little bit optimistic regarding the achievable antenna directivity, it shows many important features of cavity antennas. The plot has local maxima and minima, meaning that not just every cavity size works fine. There are some cavity sizes that provide particularly good antenna performance. These fortunate cavity sizes may achieve aperture efficiencies beyond 100%!

It is interesting to notice that the real directivity plot does not come to an end as suggested by Ehrenspeck many years ago. As shown at the end of this article, the real directivity plot has at least one additional maximum, corresponding to the recently developed "archery target" antenna [4]. Probably there are many more maxima at larger cavity diameters yet to be investigated.

Another important design parameter is the cavity height or (conductive) rim height surrounding

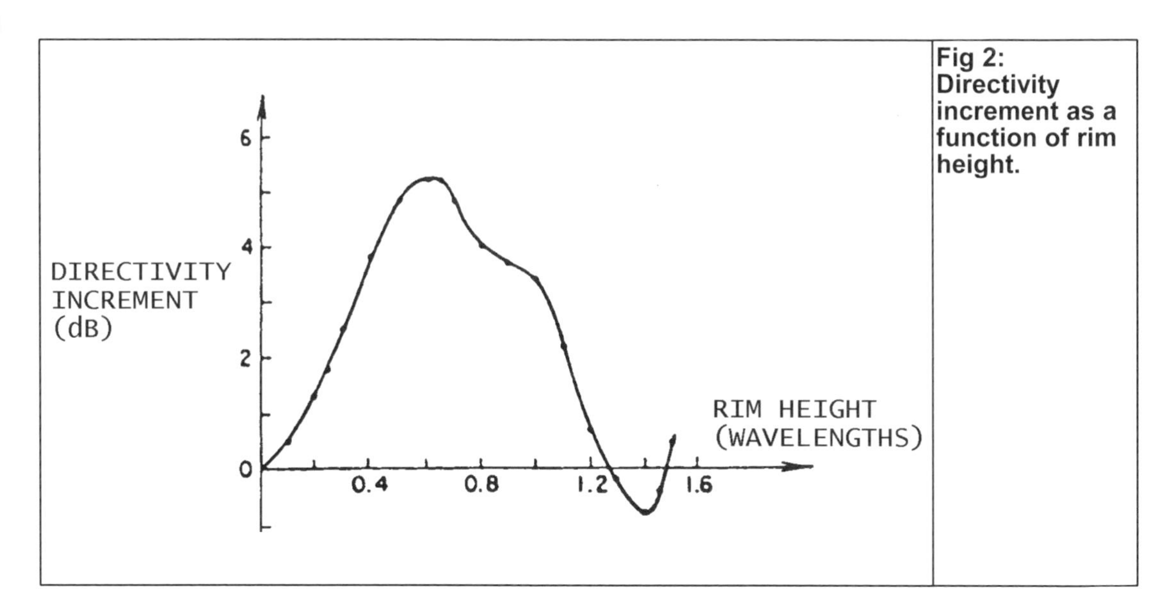

Fig 2: Directivity increment as a function of rim height.

the cavity. The maximum directivity is achieved at rim heights of one half wavelength or slightly above this value, regardless of the particular cavity diameter, as shown in Fig 2.

As shown in Fig 2, cavity antennas fill an important directivity gap between at least 7dBi and 20dBi. Since cavity antennas are simple to manufacture including weatherproofing, parabolic dishes become practical only if directivities of 22dBi or above are required.

Many practical cavity antenna designs follow directly the above mentioned directivity plots. The most important are presented in this article including practical weatherproof designs for the amateur radio frequency bands of 435MHz (70cm), 1.3GHz (23cm) and 2.3GHz (13cm). Some other important designs are omitted due to space limitations, like square cavities and cavities fed with microstrip patches.

All presented antennas were built and accurately tested many years ago at our outdoor antenna test range at the Department for Electrical Engineering of the University of Ljubljana, thanks to Mr. Stanko Gajsek. All directivity values and plots were computed from the measured E plane and H plane radiation patterns.

The measured radiation patterns shown in this article are all plotted on a 40dB logarithmic scale to have an excellent view of the side lobes and any other side effects. Please note that it is relatively easy to hide antenna design deficiencies by using a linear scale or a 20dB logarithmic scale!

All presented antennas include a transparent radome that is already part of the antenna structure. Therefore little if any additional effort is required for complete weatherproofing of these antenna designs. For terrestrial (horizontal) radio links the radiating surface is vertical, therefore rain drops, snow and ice quickly fall away if they ever stick onto the radome. Finally, the radiating surface is an equi-phase surface, meaning that a uniform coverage with ice or other foreign material does not defocus the antenna.

The presented antennas were initially used for 38.4kbps and 1.2Mbps links in the amateur packet radio network in Slovenia. Some of these antennas accumulated 15 years of continuous

Fig 3: Cup dipoles and SBFAs of CPRST:S55YCP

operation in extreme climatic environments on mountaintops. During these 15 years, rain, snow and ice never caused any link dropouts.

A typical example is the packet radio node CPRST:S55YCP installed on a mountain hut about 1840m above-sea-level and powered by solar panels. The antenna system of the latter includes a GP for 2m, two cup dipoles for 70cm, one cup dipole for 23cm, one SBFA for 23cm, one SBFA for 13cm and a webcam as shown in Fig 3.

Circular waveguide horns

General circular waveguide horn design

The simplest cavity antenna is a waveguide horn. Waveguide horns are foolproof antennas: whatever horn of whatever size and shape will always provide some useful directivity. Waveguide horns are also among the best-known cavity antennas in the amateur radio community, therefore they will be just briefly mentioned in this article.

The simplest waveguide horn is just a truncated waveguide, either of rectangular or circular cross section. Such an antenna provides a gain of about 7dBi and a main lobe width (-3dB) of around 90 degrees. These figures are useful when a broad coverage is required or to illuminate a deep (f/d=0.3 - 0.4) parabolic dish.

At UHF and lower microwave frequencies such an antenna usually includes a coax-to-waveguide transition. A practical solution is shown in Fig 4.

A quarter wavelength probe is used to excite the fundamental TE11 mode in a circular waveguide. One end of the waveguide is shorted while the other end acts as a radiating aperture. The distance between the probe and the short should be around one quarter

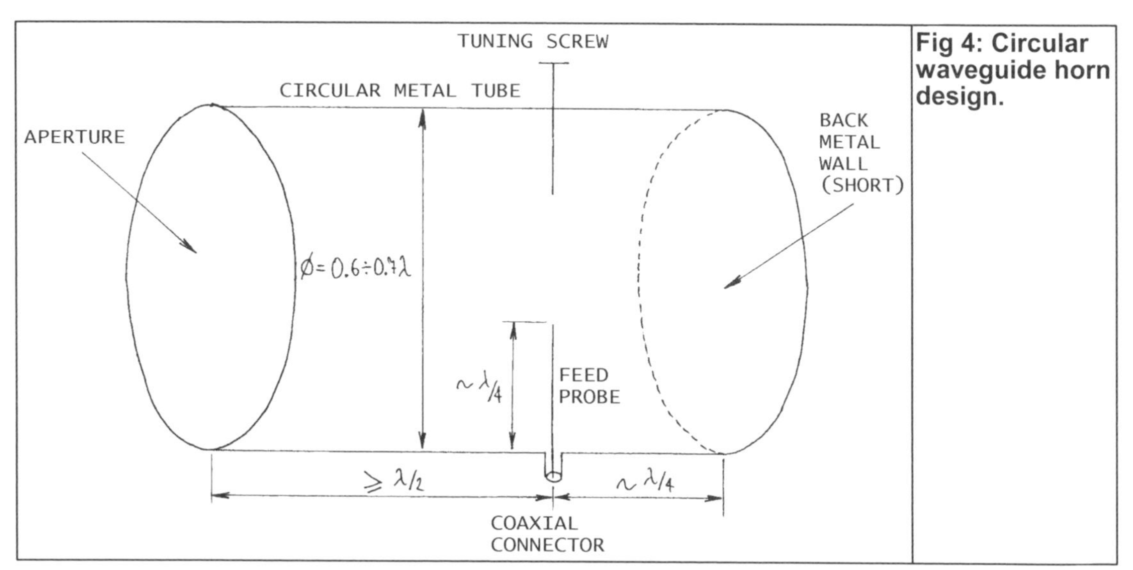

Fig 4: Circular waveguide horn design.

wavelength, to be adjusted for best impedance matching. The distance between the probe and the aperture should be at least one half wavelength to suppress higher order modes inside the waveguide. One or more tuning screws may be added to improve impedance matching. A simple feed probe generates a linearly polarised TE11 mode. Two feed probes fed in quadrature or a single feed probe and many more tuning screws inserted in a longer circular waveguide are required to obtain circular polarisation.

Practical circular waveguide horn for 23cm

Various size coffee cans usually make useful linearly polarised horns for 1.7GHz and 2.4GHz. A weatherproof waveguide horn for 1.3GHz (23cm) is a little bit too large for practical cans and will probably have to be purpose built from aluminium sheet. The required dimensions for operation around 1280MHz are shown in Fig 5.

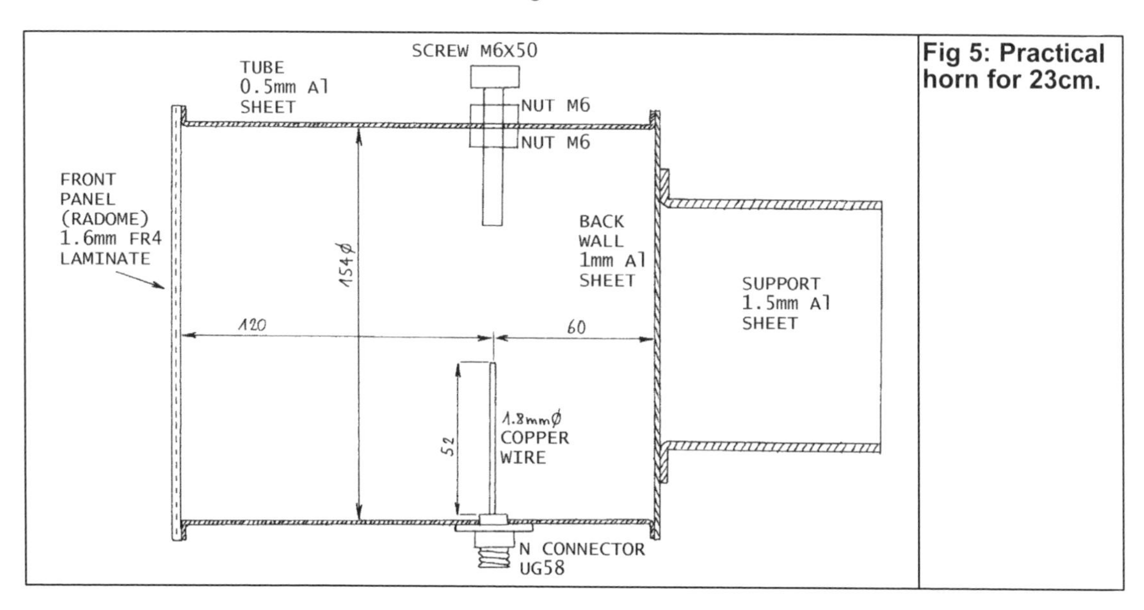

Fig 5: Practical horn for 23cm.

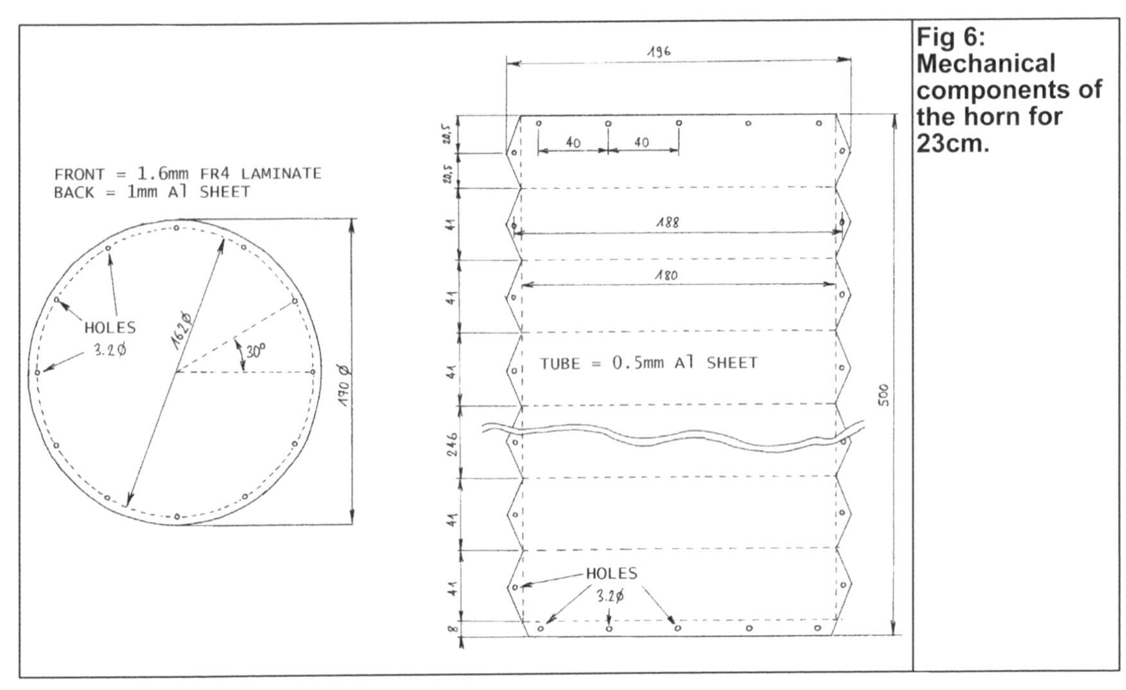

Fig 6: Mechanical components of the horn for 23cm.

The radiating aperture is covered by a disc of FR4 laminate (with any copper plating removed!) or thin plexiglass that acts as a radome. The position of the tuning screw and the length of the feed probe correspond to the given design including all dimensions and the effect of the radome. If the waveguide section is made longer, the position of the tuning screw and the length of the probe will necessarily change!

If only simple tools are available, then it makes sense to build the individual components of the horn from aluminium sheet and bolt them together with small M3 x 4 or M3 x 5 screws.

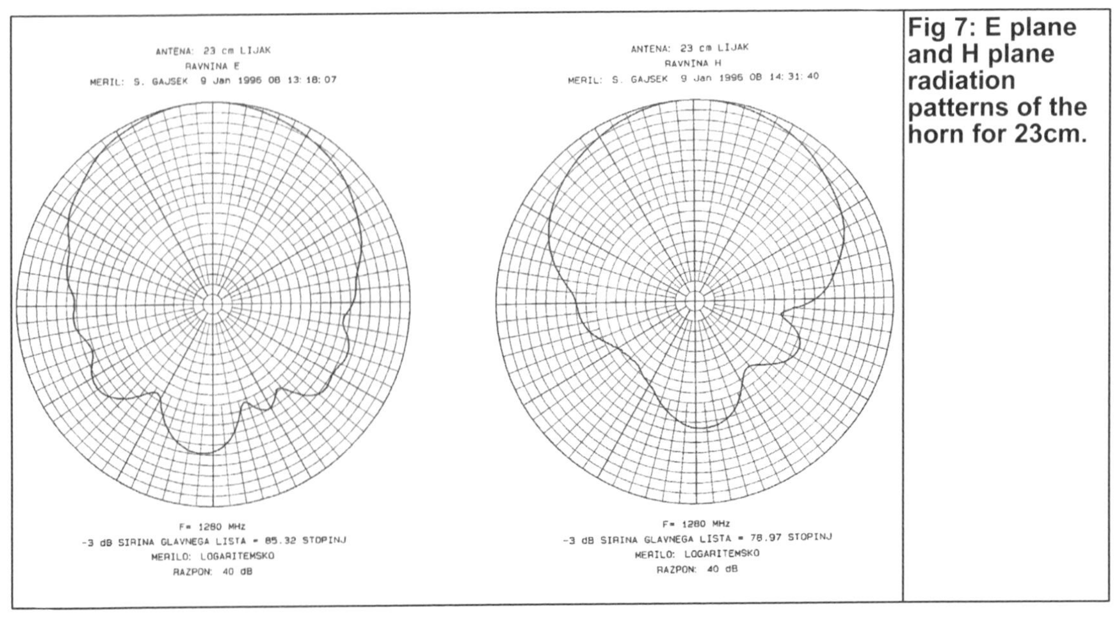

Fig 7: E plane and H plane radiation patterns of the horn for 23cm.

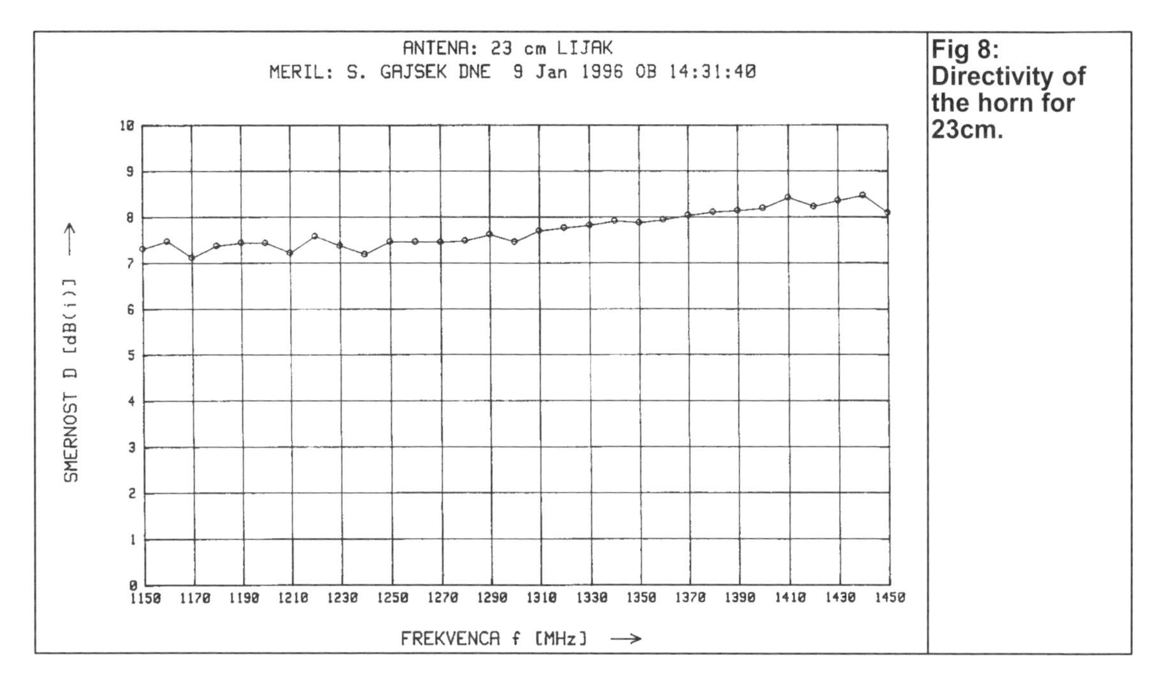

Fig 8: Directivity of the horn for 23cm.

Aluminium is a good electrical conductor providing low losses in the antenna structure and does not require any special environmental protection. The required mechanical components for the horn for 23cm are shown in Fig 6.

The antenna is first assembled together using just bolts, making all necessary adjustments like feed probe length and tuning screw position. Afterwards the antenna is disassembled so that all seams can be sealed with small amounts of silicone sealant. Finally, do not forget a venting hole or unsealed seam in the bottom part of the antenna, where any (condensation) moisture can find its way out of the antenna!

The measured E plane and H plane radiation patterns of the prototype antenna are shown in Fig 7.

The measured patterns in both planes at a number of different frequencies were used to compute the directivity as shown in Fig 8.

The operation of a simple waveguide horn is disrupted when higher order modes are excited. In the case of a simple feed probe in a circular waveguide, the first disturbing mode is the TM01 mode. The latter causes an unsymmetrical illumination of the aperture resulting in a squint of the direction of radiation.

The appearance of higher order waveguide modes means that not every coffee can makes a useful antenna for the desired frequency range! The prototype horn for 23cm displays a large squint of the main lobe due to higher order modes at 1450MHz as shown in Fig 9.

Cup dipole

Basic design of a cup dipole

In order to increase the gain of a waveguide horn, the size of the aperture has to be increased without exciting too many higher order waveguide modes. A rather simple solution is to make a

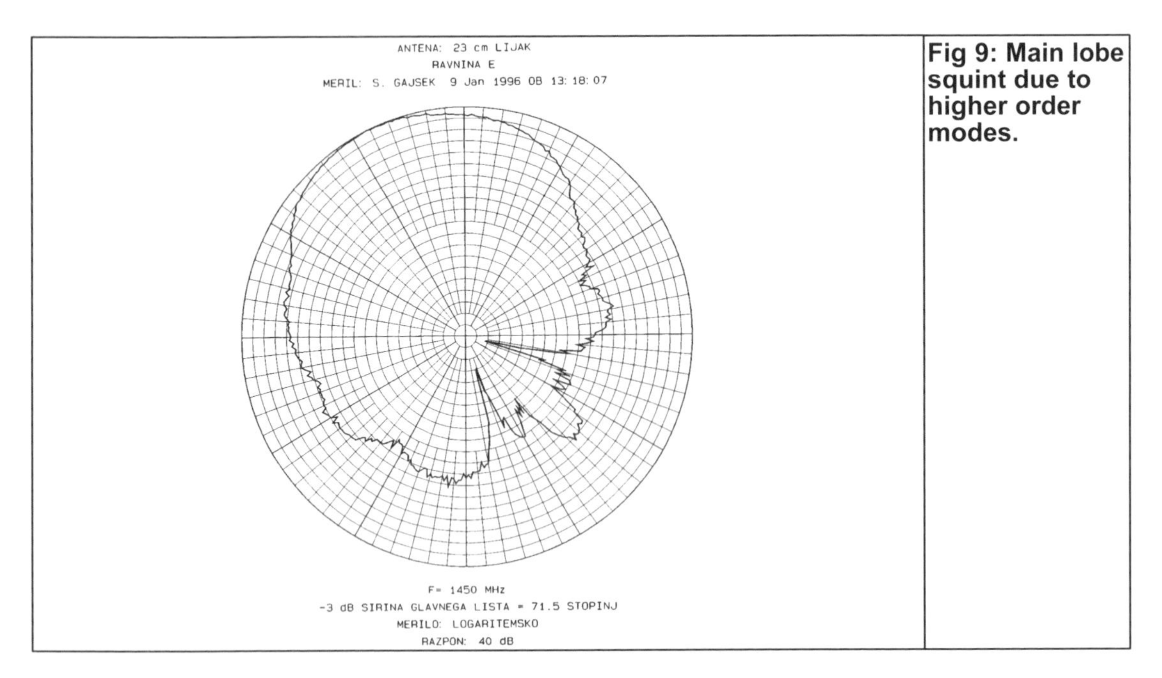

Fig 9: Main lobe squint due to higher order modes.

smooth transition from the waveguide to the larger aperture in the form of a pyramidal or conical horn. Such a solution becomes impractical at lower microwave frequencies and UHF, where the size of the horn is too large.

An alternative solution is to avoid exciting unwanted modes already at the transition from an arbitrary TEM feed line to the waveguide. Replacing a simple feed probe with a symmetrical half wave dipole avoids exciting the unwanted TM01 mode while exciting the desired TE11 mode. Such a solution is called a cup dipole and is represented in Fig 10.

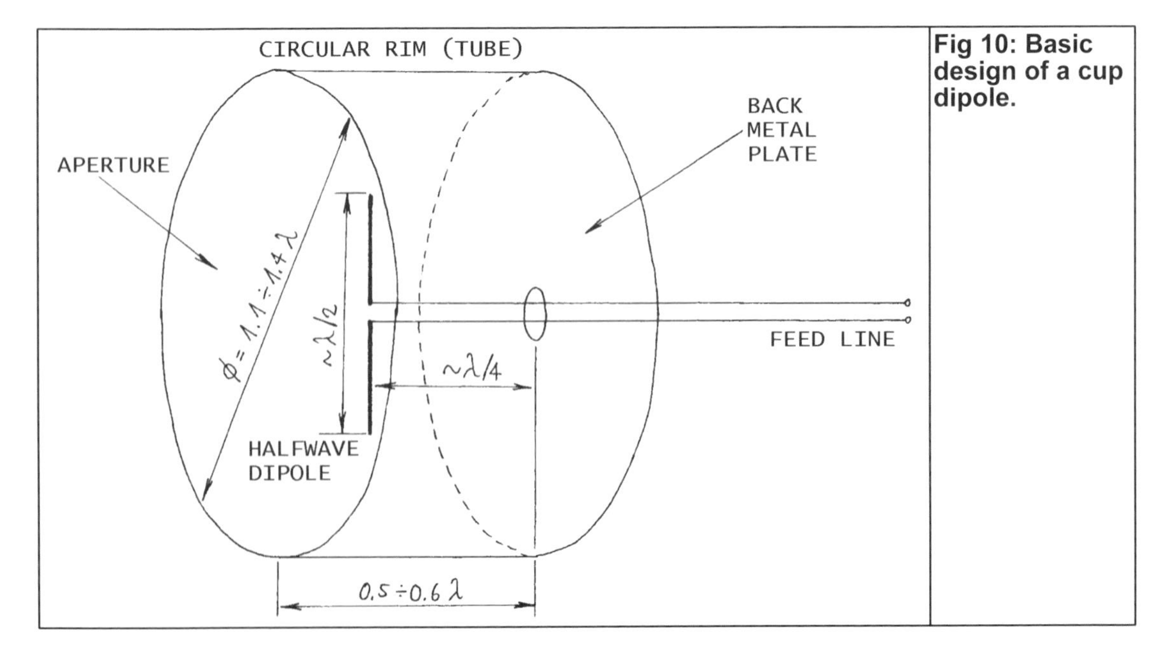

Fig 10: Basic design of a cup dipole.

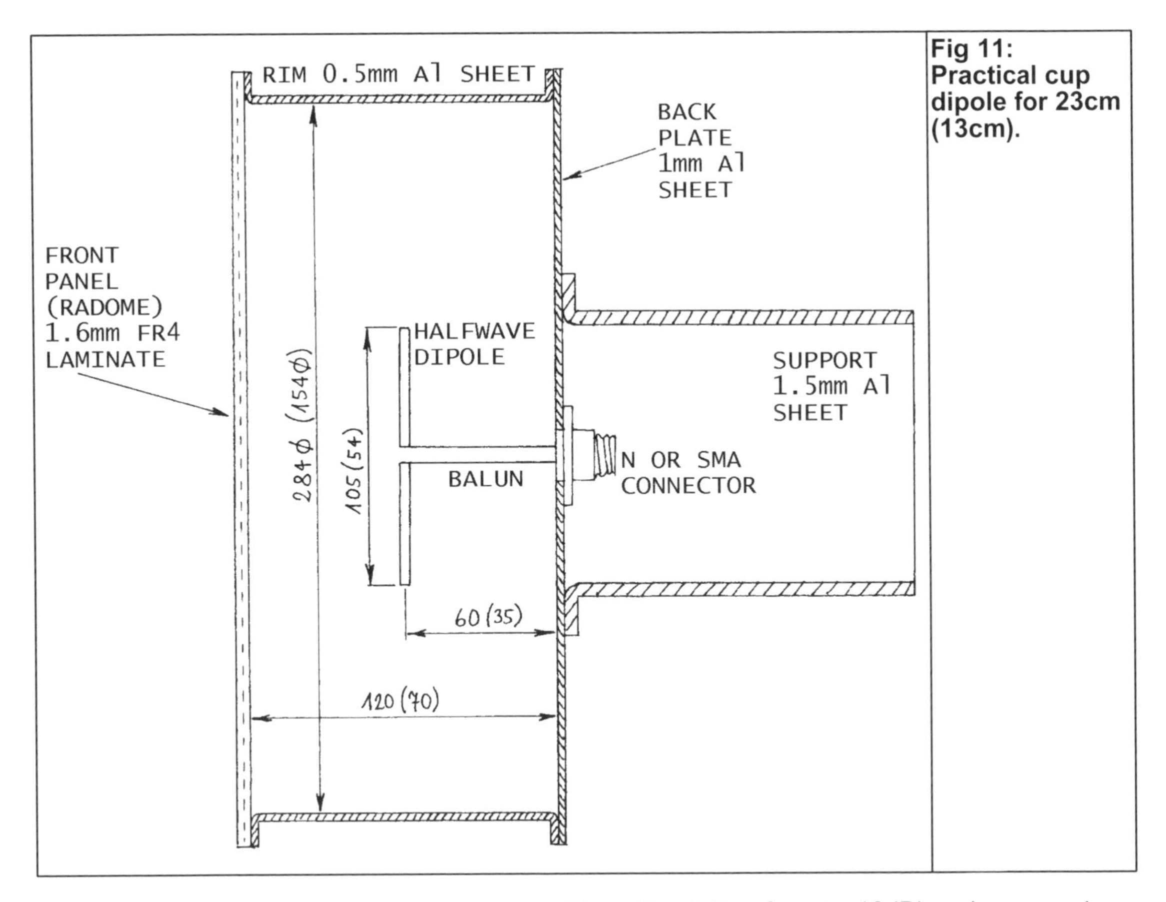

Fig 11: Practical cup dipole for 23cm (13cm).

A cup dipole is a really compact antenna with a directivity of up to 12dBi and a very clean radiation pattern with very weak side lobes. The directivity achieves its maximum at the second peak on the Ehrenspeck's diagram [3]. Again, most of the metallic antenna structure can be used as a radome at the same time and just the radiating aperture needs to be covered with a transparent cover.

A cup dipole provides a -3dB beam width of about 50 degrees. Besides operating as a standalone antenna, a cup dipole also makes an excellent feed for a shallow (f/d=0.6 - 0.7) parabolic dish.

Cup dipole for 23cm

The construction of a practical cup dipole for 23cm (13cm) is shown in Fig 11.

If only simple tools are available, then it makes sense to build the individual components of the cup dipole from aluminium sheet and bolt them together with small M3 x 4 or M3 x 5 screws. Aluminium is a good electrical conductor providing low losses in the antenna structure and does not require any special environmental protection. The required mechanical components for the cup dipole for 23cm are shown in Fig 12.

The front cover may be quite thick FR4 laminate or plexiglass, since a dielectric plate in this position actually improves the performance of a cup dipole.

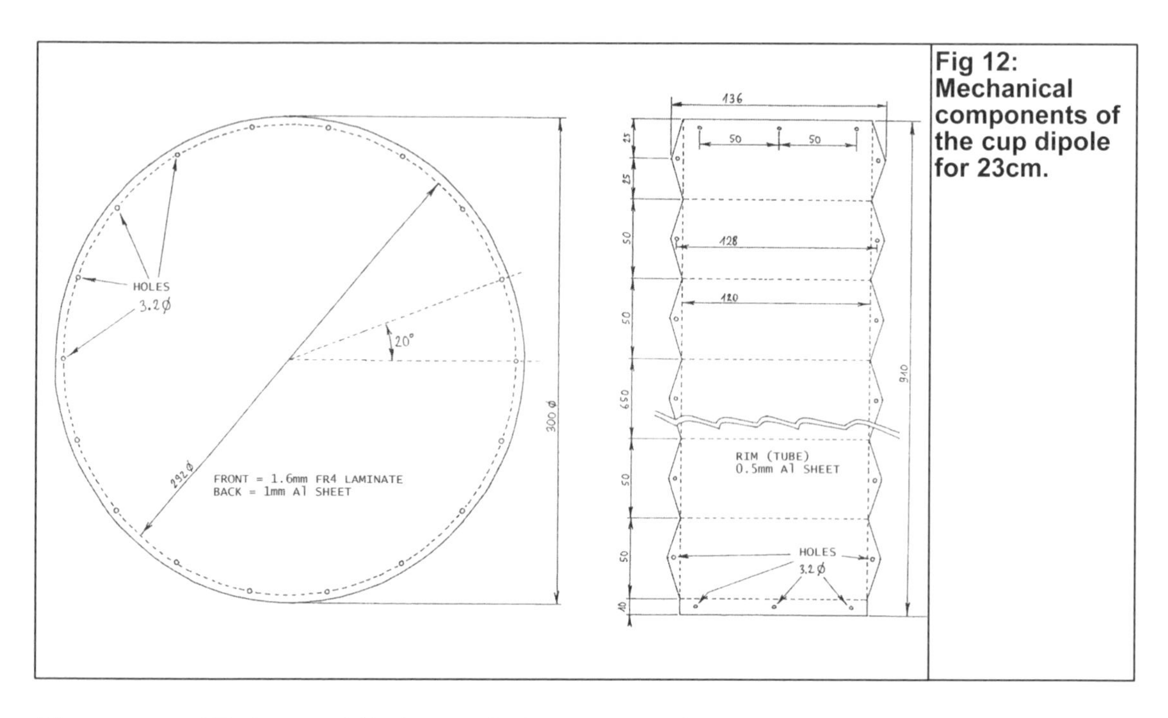

Fig 12: Mechanical components of the cup dipole for 23cm.

The measured E plane and H plane radiation patterns of the prototype cup dipole are shown in Fig 13.

The measured patterns in both planes at a number of different frequencies were used to compute the directivity as shown in Fig 14.

The kink in the directivity curve will be explained later together with the same effect observed with the prototype for 13cm.

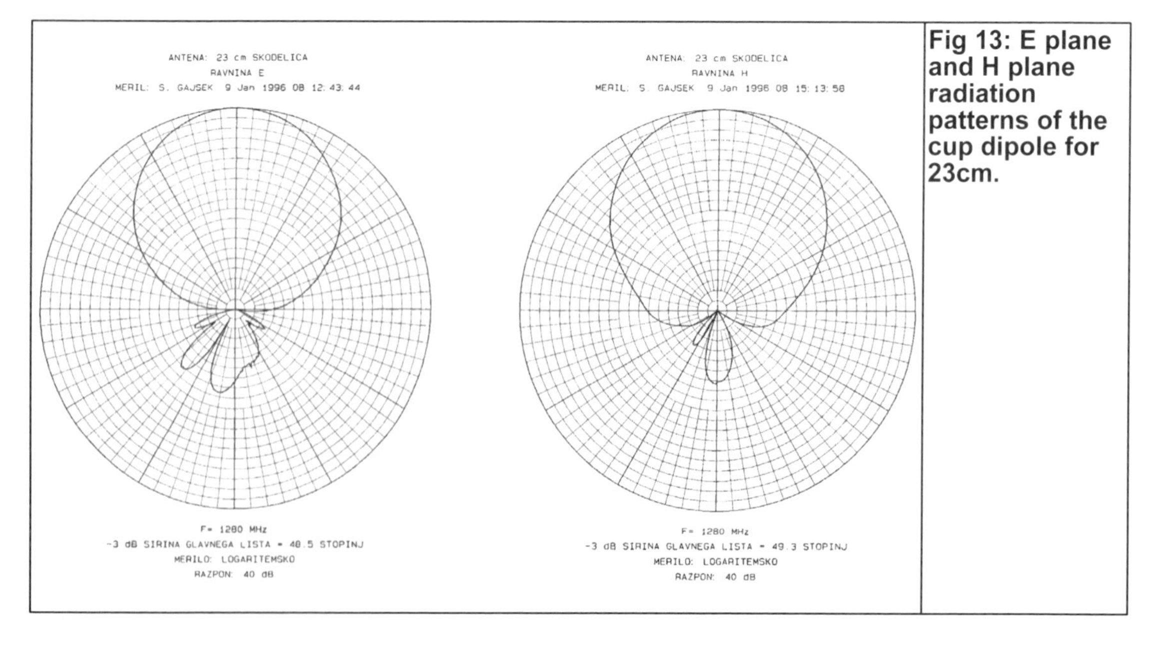

Fig 13: E plane and H plane radiation patterns of the cup dipole for 23cm.

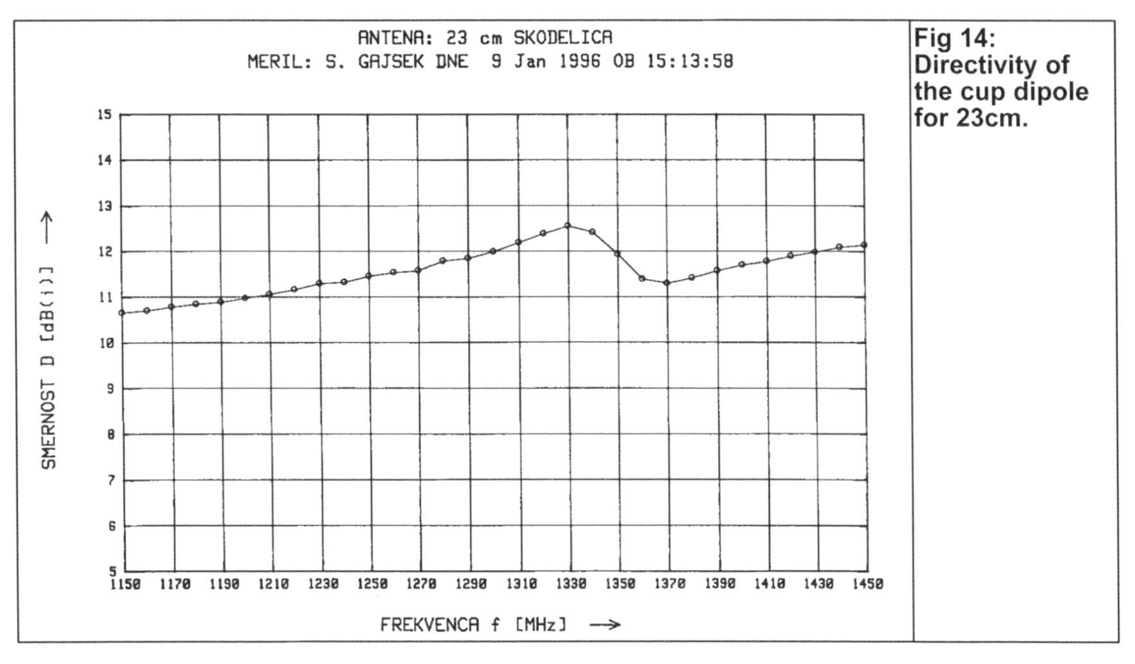

Fig 14: Directivity of the cup dipole for 23cm.

Cup dipole for 13cm

The design of the cup dipole can be easily scaled to the 13cm band. The required mechanical components for the cup dipole for 13cm are shown in Fig 15.

The antenna is first assembled together using just bolts, making all necessary adjustments and check-outs. Afterwards the antenna is disassembled so that all seams can be sealed with small

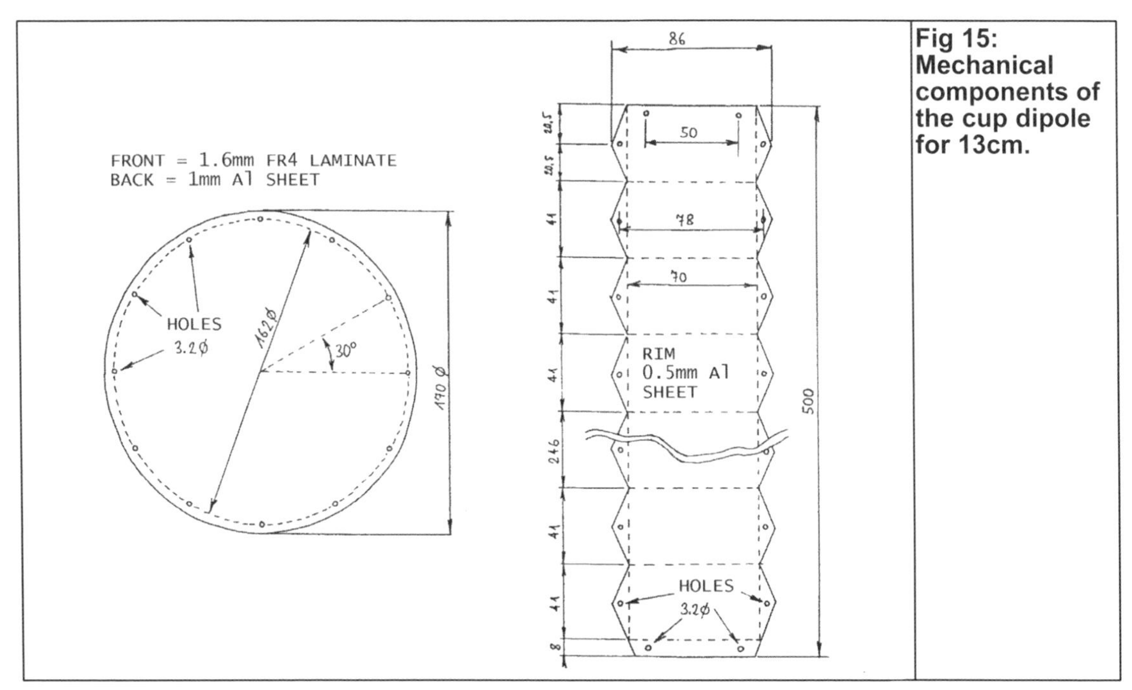

Fig 15: Mechanical components of the cup dipole for 13cm.

Fig 16: Cup dipole for 13cm.

amounts of silicone sealant. Finally, do not forget a venting hole or unsealed seam in the bottom part of the antenna, where any (condensation) moisture can find its way out of the antenna!

A home-made cup dipole for 13cm that already provided several years of outdoor service is shown in Fig 16.

The measured E plane and H plane radiation patterns of the prototype cup dipole for 13cm are shown in Fig 17.

The measured patterns in both planes at a number of different frequencies were used to compute the directivity as shown in Fig 18.

As the frequency increases, the directivity plots of both cup dipoles for 23cm and 13cm include a kink. A further explanation of what is happening is given by the two E plane radiation patterns for both investigated antennas: the 23cm cup dipole at 1370MHz and the 13cm cup dipole at 2480MHz: shown in Fig 19.

Both radiation patterns are badly corrupted due to the appearance of higher order symmetrical waveguide modes. These are excited by a perfectly symmetrical half wave dipole and are no longer suppressed by the waveguide. The size and directivity of a cup dipole therefore has a practical upper limit.

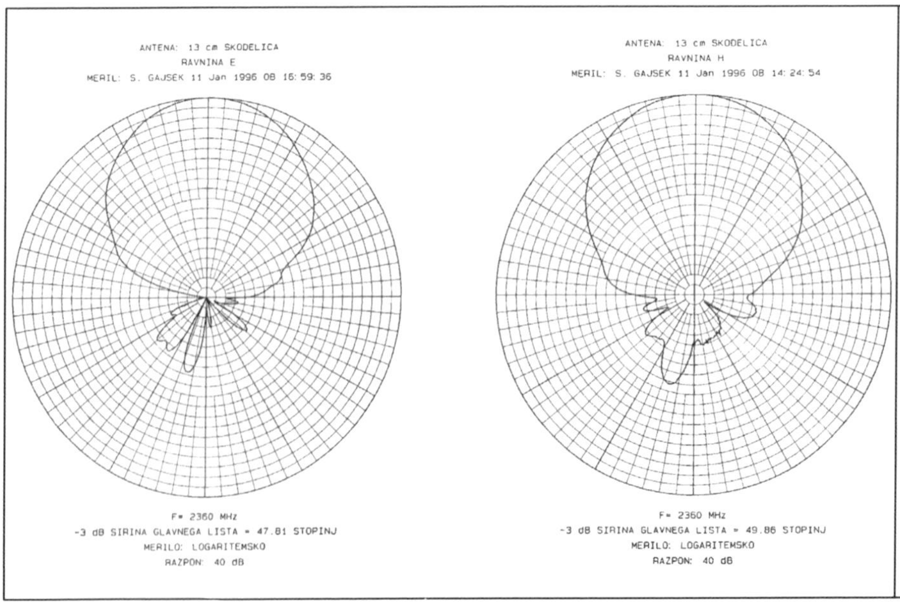

Fig 17: E plane and H plane radiation patterns of the cup dipole for 13cm.

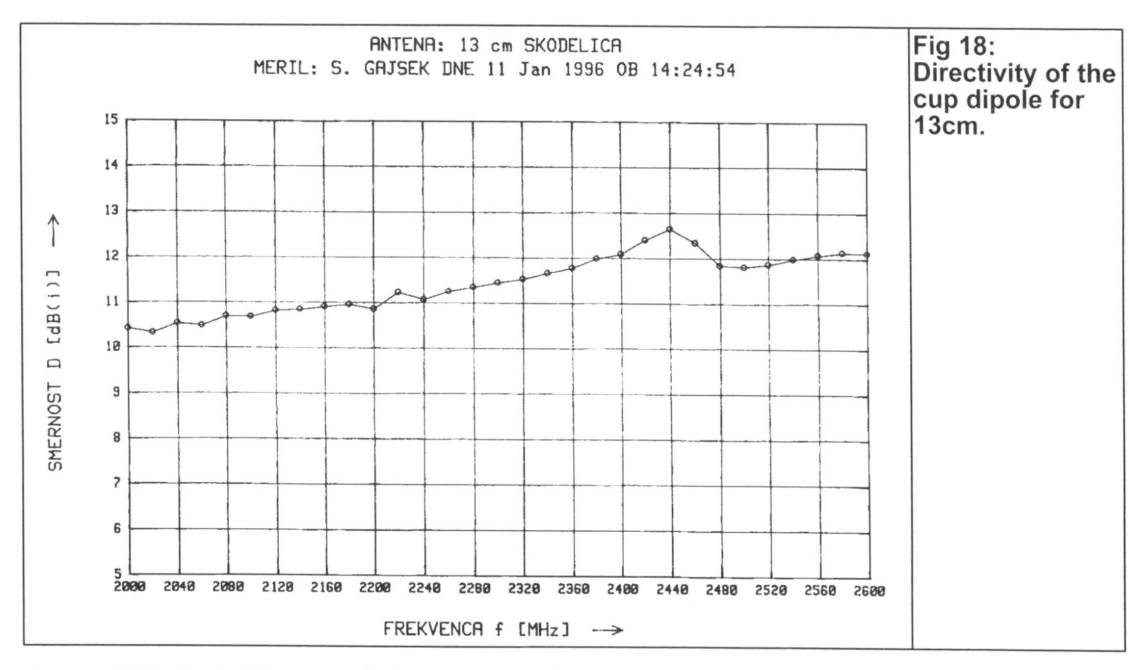

Fig 18: Directivity of the cup dipole for 13cm.

Beyond this limit different solutions are required to control the illumination of the aperture to obtain even narrower radiation beams and higher values of directivity.

Cup dipole for 70cm

The opposite happens at UHF and lower frequencies: all presented cavity antennas are physically too large to be practical. At low frequencies, a good hint is to look at the design of very compact coaxial-to-waveguide transitions. A practical solution in the 70cm band is a down-scaled cup dipole as shown in Fig 20.

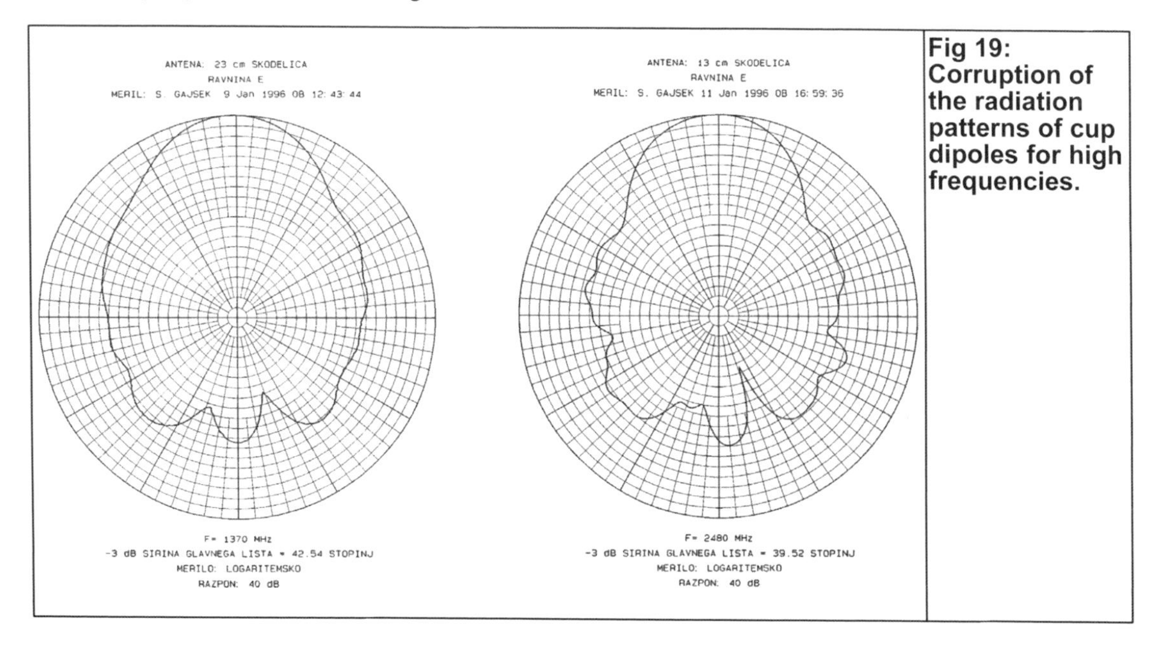

Fig 19: Corruption of the radiation patterns of cup dipoles for high frequencies.

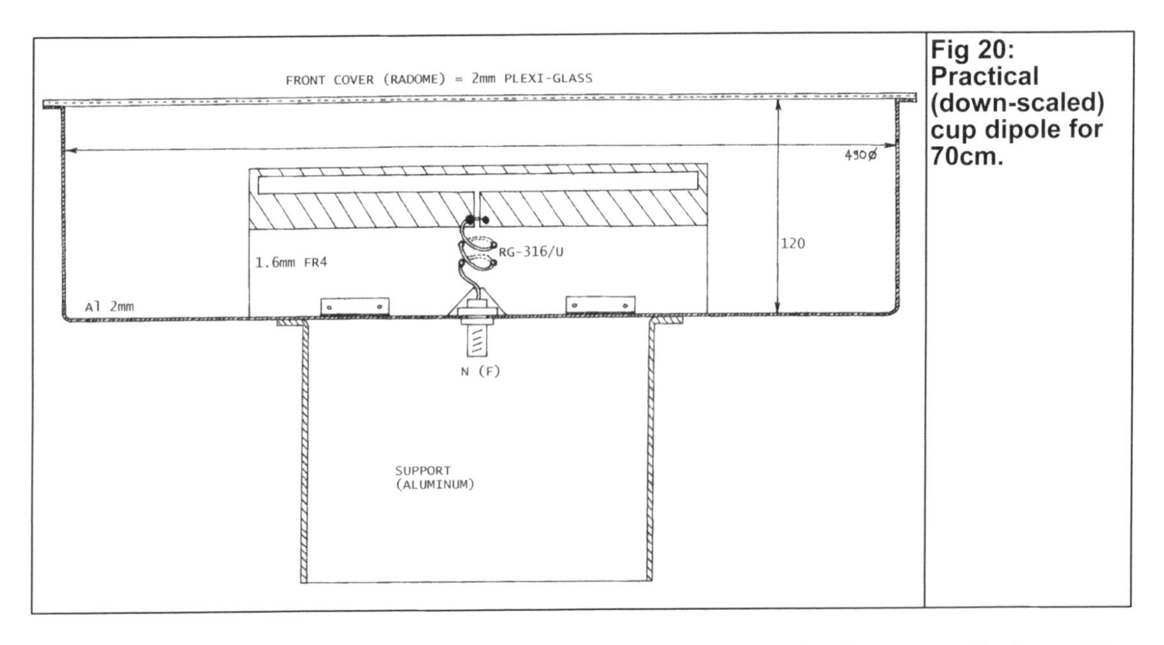

Fig 20: Practical (down-scaled) cup dipole for 70cm.

The aperture of this down-scaled cup dipole corresponds to a simple waveguide horn. The expected radiation pattern is therefore similar to the much longer waveguide horn and the expected directivity is 8dBi or less. 8dBi may not seem much, but remember that this figure is achieved with a compact and weatherproof antenna at a relatively low frequency!

Since the half wave dipole is installed rather close to the cavity wall, its expected radiating impedance will be very low. A folded dipole is therefore used for impedance transformation. The folded dipole is built on a printed circuit board and requires a balanced 50Ω feed.

The balun is simply a quarter wavelength piece of RG-316/U thin Teflon dielectric coaxial cable forming a two-turn coil. The dipole is built on a piece of 1.6mm thick FR4 laminate with 35um or thicker copper cladding. No etching is usually required. The copper cladding is marked with a

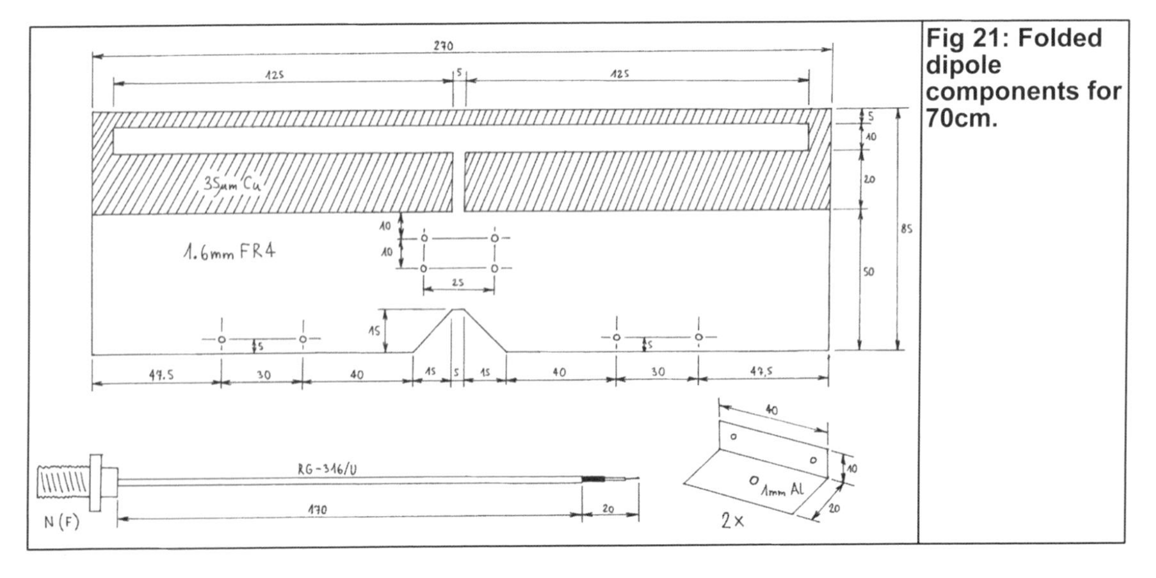

Fig 21: Folded dipole components for 70cm.

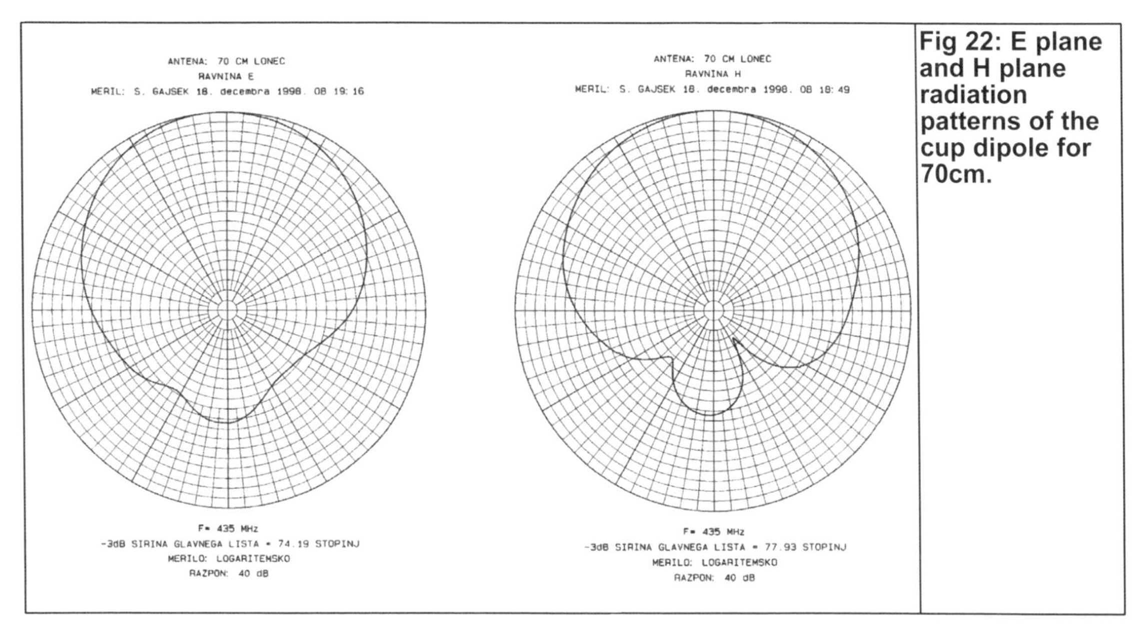

Fig 22: E plane and H plane radiation patterns of the cup dipole for 70cm.

sharp tip and the unnecessary copper foil is simply peeled off.

All of the folded dipole components are shown in Fig 21, including two aluminium brackets used to bolt the dipole to the cavity wall.

The measured E plane and H plane radiation patterns of the prototype cup dipole for 70cm are shown in Fig 22.

The measured patterns in both planes at a number of different frequencies were used to compute the directivity as shown in Fig 23.

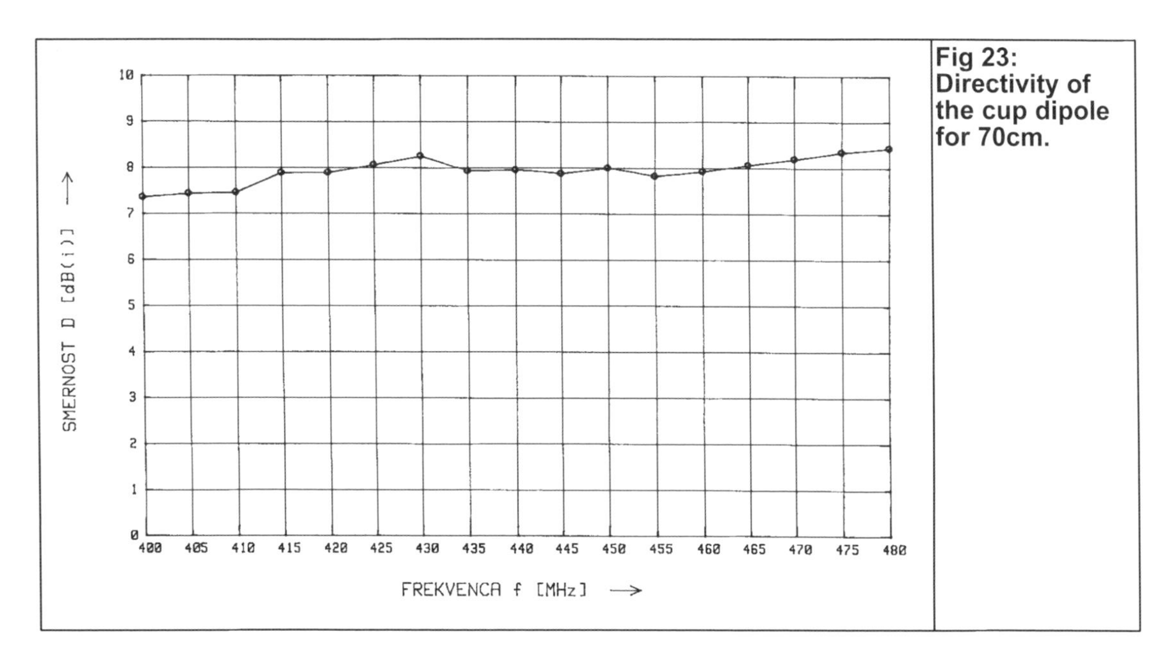

Fig 23: Directivity of the cup dipole for 70cm.

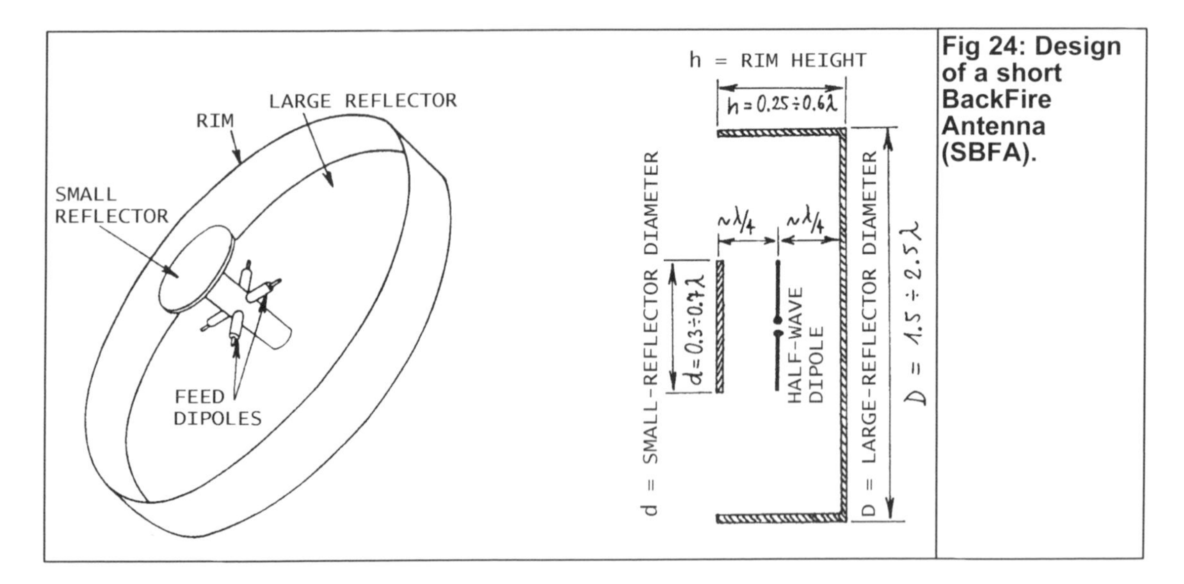

Fig 24: Design of a short BackFire Antenna (SBFA).

Short backfire antenna (SBFA)

Design of a short backfire antenna (SBFA)

As the diameter of the cup dipole cavity increases beyond 1.4 wavelengths, additional symmetrical circular waveguide modes are excited by the dipole feed and propagated in the waveguide. These modes spoil the aperture illumination, increase the side lobe levels and decrease the antenna directivity. In order to build even larger cavity antennas, some means of controlling the amplitude and phase of all contributing modes has to be introduced.

The most popular solution to control the amplitude and phase of symmetrical modes inside a circular waveguide is to introduce an additional circular plate in the aperture plane. This plate is called the small reflector while the cavity is called the large reflector. Together with one or more feed dipoles, these two reflectors form a very efficient cavity antenna called the short backfire antenna or SBFA as shown in Fig 24.

The large reflector of a SBFA has a diameter D of up to 2.5 wavelengths while the small reflector has a diameter of about one half wavelength or slightly more. The rim height h is usually around one half wavelength. The directivity of such a simple antenna exceeds 16dBi with an aperture efficiency close to 100% corresponding to the third peak on the Ehrenspeck's diagram [3]. Again, most of the metal structure can be used as a radome and just the radiating aperture needs to be covered with a transparent cover.

A SBFA provides a -3dB beam width of about 30 degrees. The whole antenna structure is not critical and is able to operate over bandwidths of more than 10% of the central frequency. Since the SBFA reflector structure is rotationally symmetrical, the polarisation only depends on the feed. Two feed dipoles may be used for dual polarisation or circular polarisation.

SBFA for 23cm

The construction of a practical SBFA for 23cm (13cm) is shown in Fig 25

The front panel (radome) may act as a support for the small reflector, thus considerably

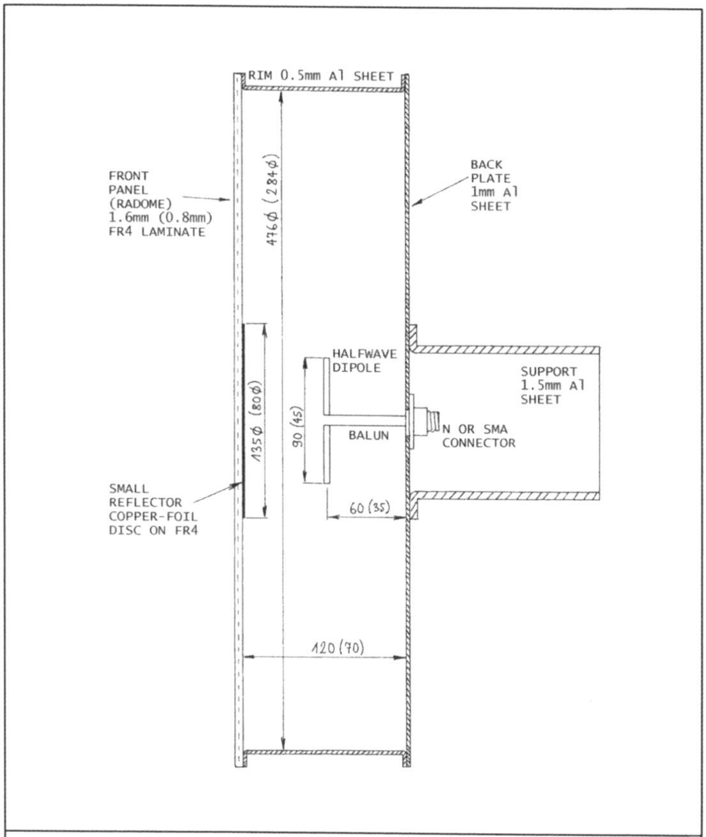

Fig 25: Practical SBFA for 23cm (13cm).

simplifying the mechanical design of the antenna. The radome has some measurable effect on the SBFA and any dielectric should not be too thick. If only simple tools are available, then it makes sense to build the metal components of the SBFA from aluminium sheet and bolt them together with small M3 x 4 or M3 x 5 screws.

The front panel (radome) is a disc of 1.6mm (0.8mm for 13cm) FR4 laminate. The small reflector is a disc of copper foil (35um or thicker cladding). No etching is usually required. The copper cladding is marked with a sharp tip and the unnecessary copper foil is simply peeled off. The required mechanical components for the SBFA for 23cm are shown in Fig 26.

The measured E plane and H plane radiation patterns of the prototype cup dipole are shown in Fig 27.

The measured patterns in both

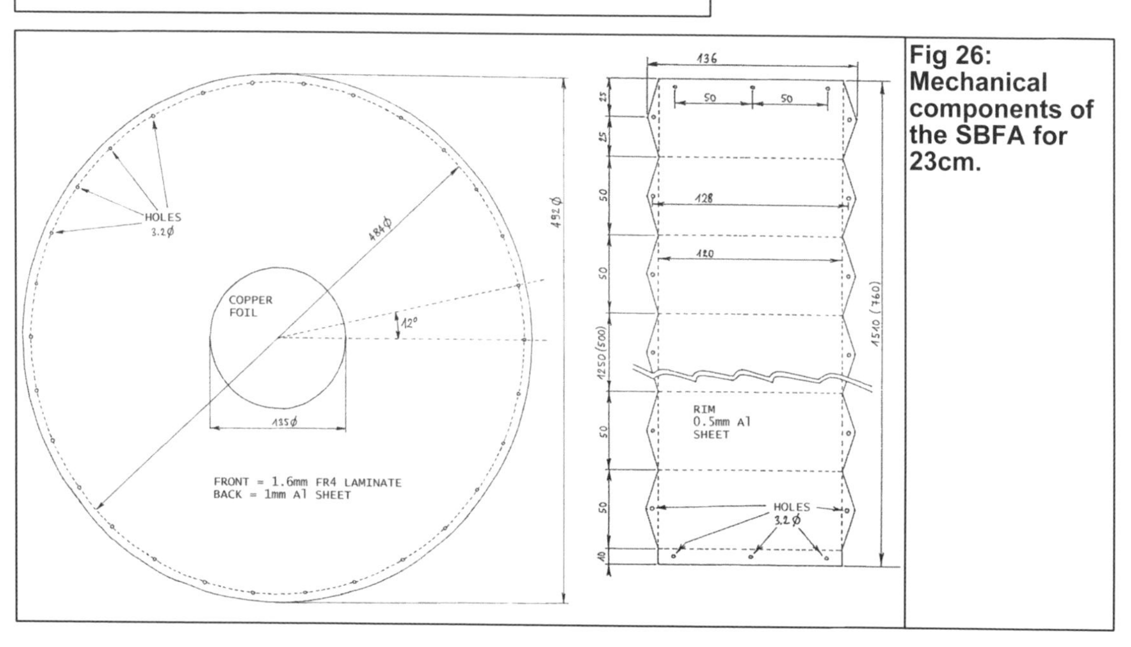

Fig 26: Mechanical components of the SBFA for 23cm.

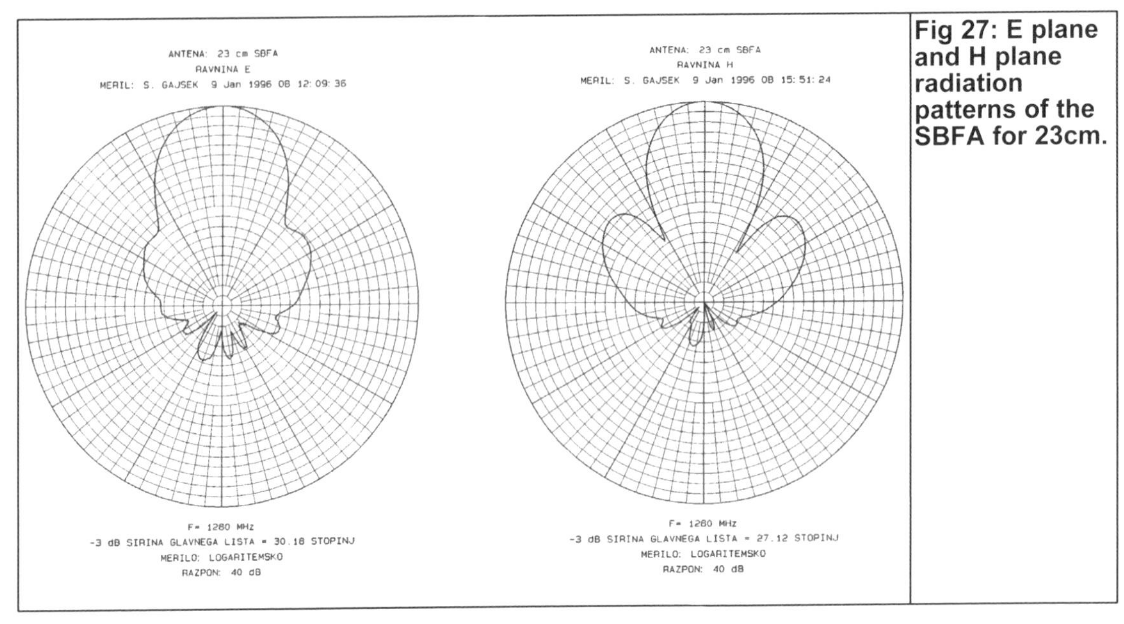

Fig 27: E plane and H plane radiation patterns of the SBFA for 23cm.

planes at a number of different frequencies were used to compute the directivity as shown in Fig 28.

SBFA for 13cm

The design of the SBFA can be easily scaled to the 13cm band. Since the SBFA is sensitive to the radome thickness, the latter has to be scaled to the higher frequency as well! The required mechanical components for the SBFA for 13cm including the radome from 0.8mm thick FR4 laminate are shown in Fig 29.

The antenna is first assembled together using just bolts, making all necessary adjustments and

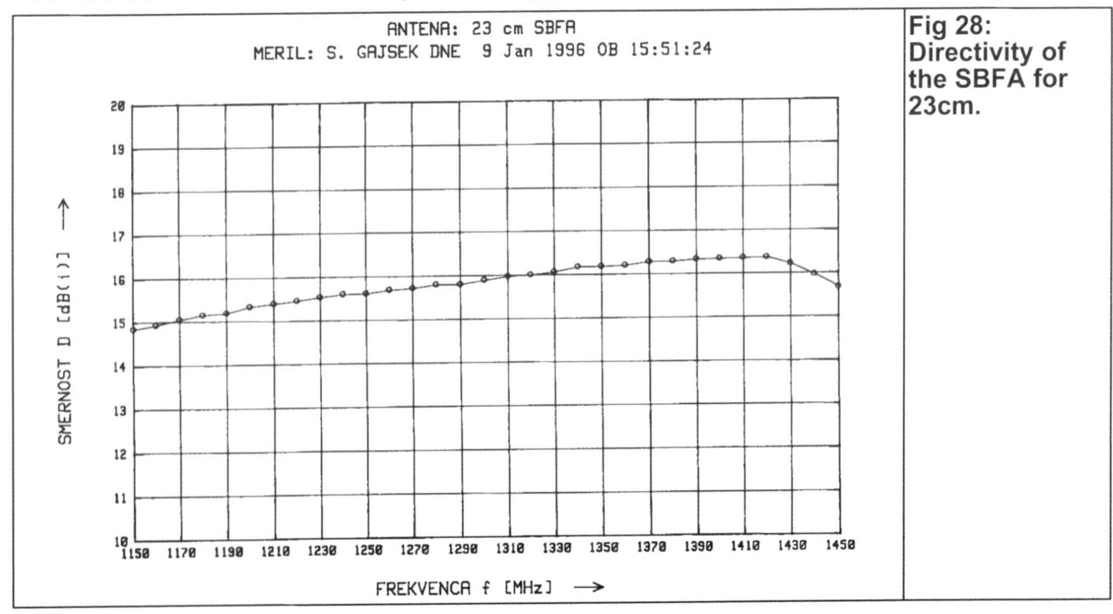

Fig 28: Directivity of the SBFA for 23cm.

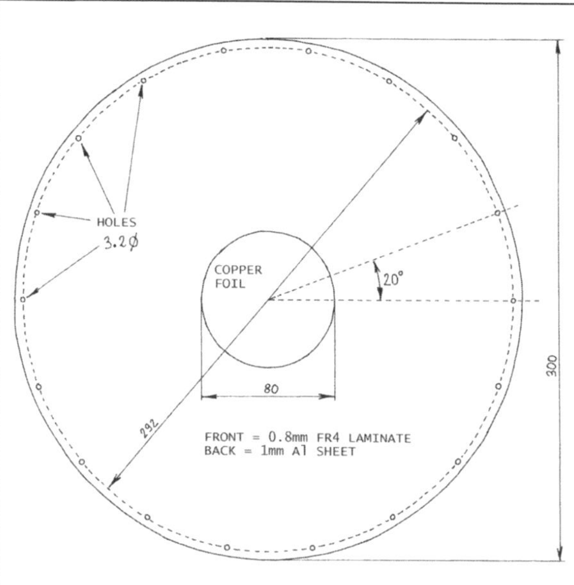

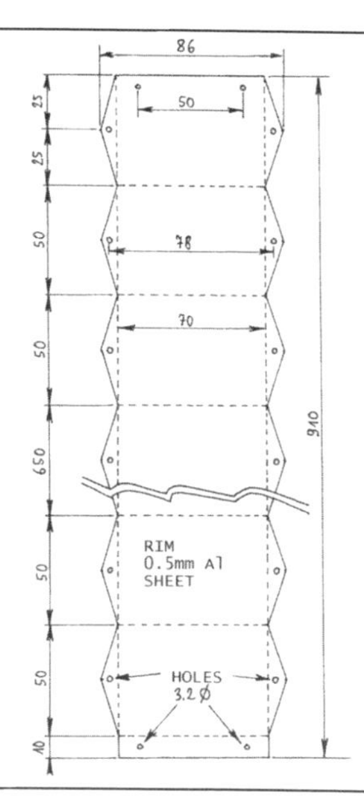

Fig 29: Mechanical components of the SBFA for 13cm.

Fig 30: Homemade SBFA for 13cm.

checkouts. Afterwards the antenna is disassembled so that all seams can be sealed with small amounts of silicone sealant. Finally, do not forget a venting hole or unsealed seam in the bottom part of the antenna, where any (condensation) moisture can find its way out of the antenna! A homemade SBFA for 13cm is shown in Fig 30.

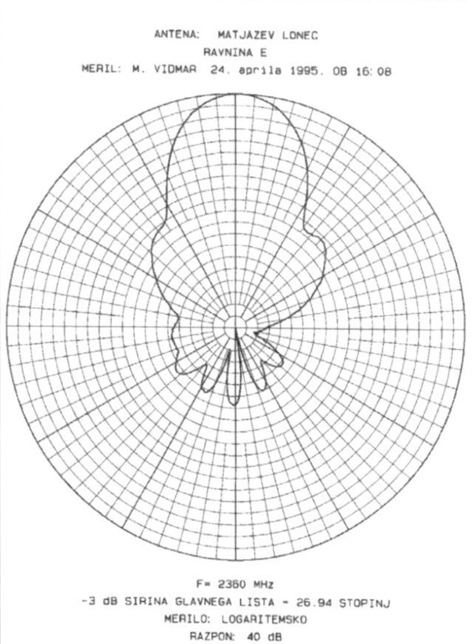

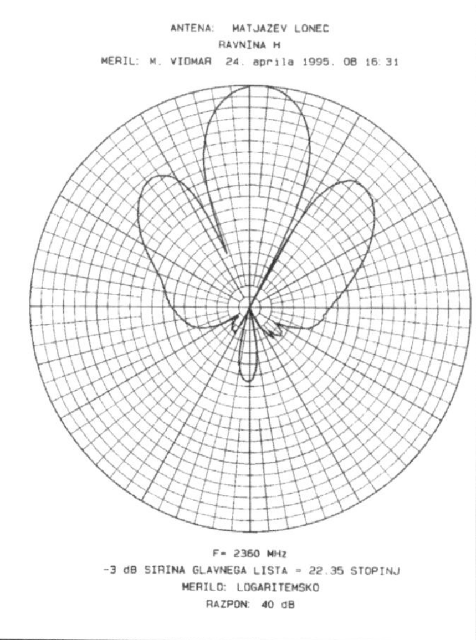

Fig 31: E plane and H plane radiation patterns of the SBFA for 13cm.

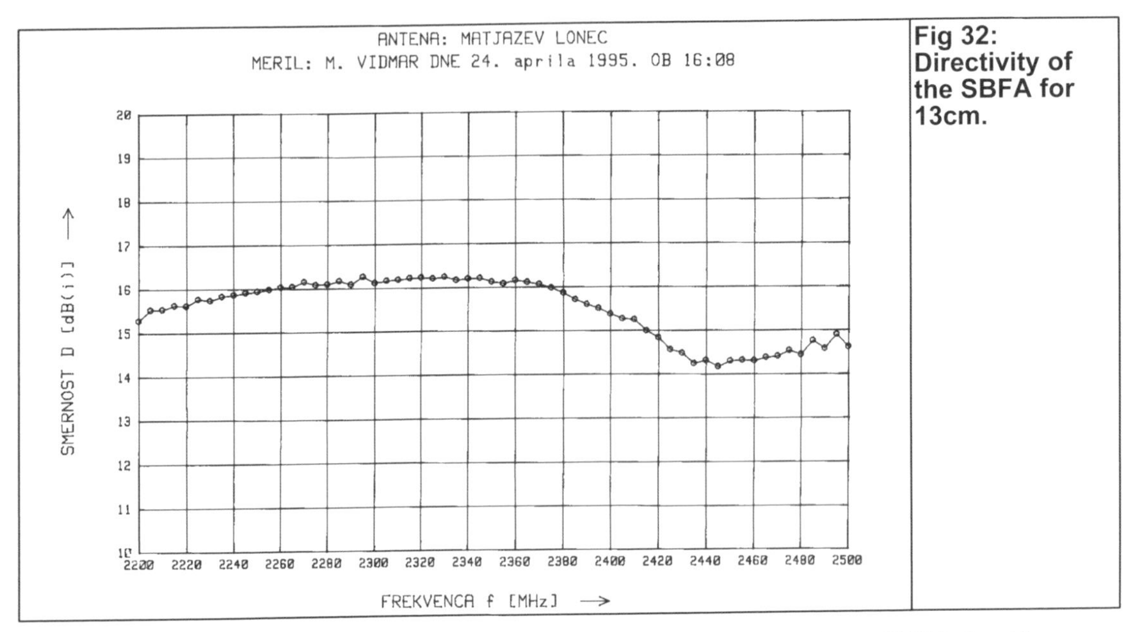

Fig 32: Directivity of the SBFA for 13cm.

The measured E plane and H plane radiation patterns of the prototype SBFA for 13cm are shown in Fig 31.

The measured patterns in both planes at a number of different frequencies were used to compute the directivity as shown in Fig 32.

As the frequency increases, the H plane side lobes of both SBFAs for 23cm and 13cm increase. As the diameter of the antenna D exceeds 2.5 wavelengths, the side lobes become so large that the directivity of the antenna starts decreasing.

Feeds for SBFAs and cup dipoles

A half wave dipole is the simplest way to feed cup dipole or SBFA cavities. Microstrip patch antennas or rectangular or circular metal waveguides can also be used. A half wave dipole requires a balun to be fed with standard 50Ω coaxial cable.

Besides the directivity bandwidth of a cup dipole or SBFA, the gain bandwidth of these antennas is also limited by the impedance matching of the feed. In the case of a SBFA, the feed is enclosed between the large and small reflectors, resulting in a low radiation impedance and corresponding sharp resonance. The impedance matching bandwidth is likely much narrower than the bandwidth of operation of the SBFA cavity.

An impedance mismatch results in a loss of antenna gain. In the case of microwave cavity antennas, this is the only significant loss mechanism, since the electrical efficiency of the cavities themselves is close to unity. A good estimate for the antenna gain is therefore just subtracting the impedance mismatch loss from the directivity.

The antenna bandwidth can be improved by a broadband feed dipole. One possible solution is to build the feed dipole from semi-rigid coaxial cable and use its internal conductor as a reactive load to broaden the impedance matching bandwidth. The same type of semi-rigid cable can also be used for the balun including a dummy arm. The wiring of such a dipole and corresponding balun is shown in Fig 33.

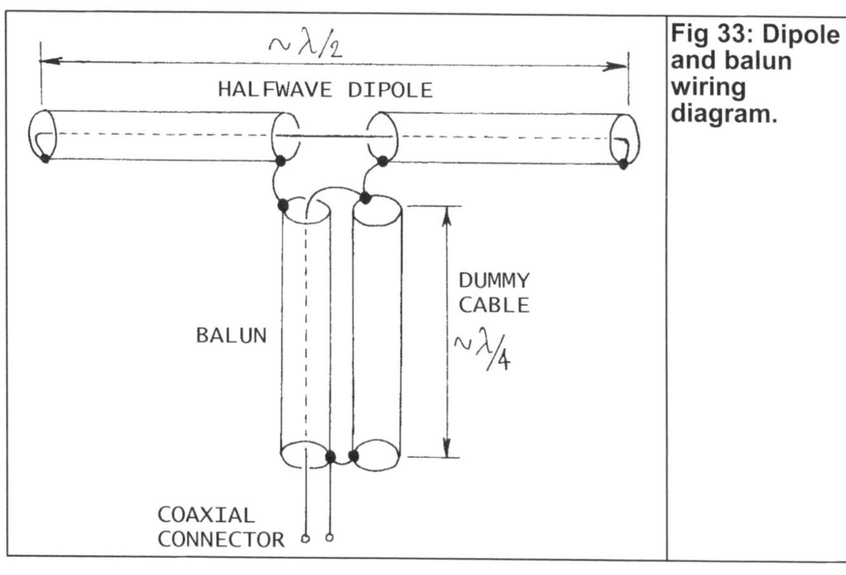

Fig 33: Dipole and balun wiring diagram.

A practical solution is to build the feed dipoles and corresponding baluns from UT-141 semi-rigid cable (outer diameter about 3.6mm) in the 23cm frequency band and from UT-085 semi-rigid cable (outer diameter about 2.2mm) in the 13cm band as shown in Fig 34.

A suitable coaxial connector for semi-rigid cable should be selected first. N connectors may be useful in the 23cm band. Smaller SMA or TNC connectors may be used in the 13cm band. While soldering semi-rigid coaxial cables one needs to take into account the thermal expansion of their Teflon dielectric!

The dipole is made by cutting the specified length "A" of semi-rigid cable. Then "D" millimetres of outer conductor and Teflon dielectric are removed at both ends. After this operation the outer conductor copper tube is carefully cut in the centre and both parts are pulled away to form the gap "C". Finally both dipole ends are filled with solder to connect the centre and outer conductors.

The published 23cm SBFA with the described feed achieves a return loss better than -10dB over the whole 1240 - 1300MHz frequency band. The 13cm SBFA design has a slightly higher

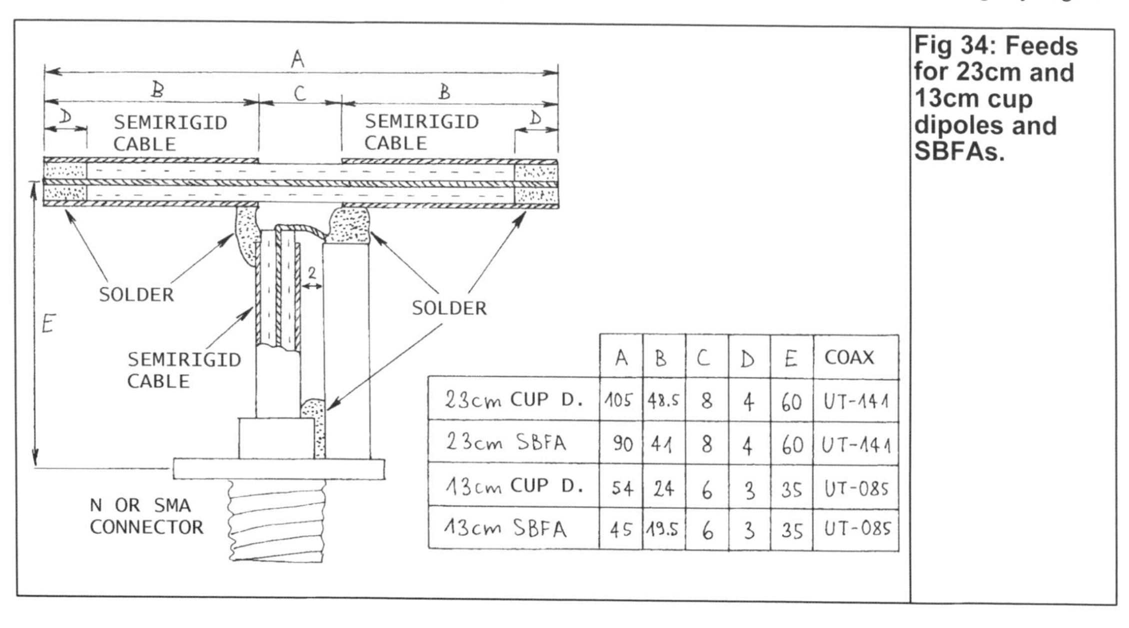

	A	B	C	D	E	COAX
23cm CUP D.	105	48.5	8	4	60	UT-141
23cm SBFA	90	41	8	4	60	UT-141
13cm CUP D.	54	24	6	3	35	UT-085
13cm SBFA	45	19.5	6	3	35	UT-085

Fig 34: Feeds for 23cm and 13cm cup dipoles and SBFAs.

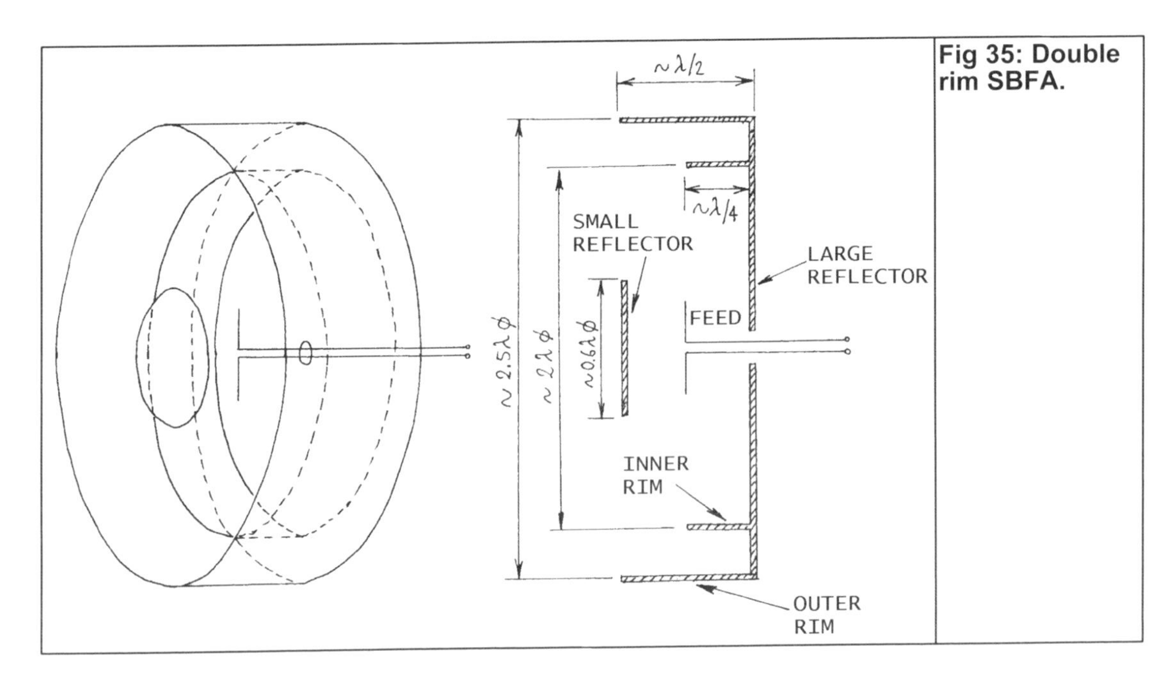

Fig 35: Double rim SBFA.

directivity resulting in a narrower bandwidth. The 13cm SBFA with the described feed achieves a return loss of -10dB over the frequency band 2300 - 2360MHz.

The cavities of cup dipoles represent a different load for the feed dipoles; therefore the dimensions of the feed dipoles are necessarily different from those used for the SBFAs. The impedance matching bandwidth of cup dipoles is much broader than SBFAs and a return loss of -15dB can usually be achieved.

Beyond the SBFA directivity

Double rim SBFA

The simplicity, efficiency and performance of the short backfire antenna suggests looking for similar antenna solutions also for a cavity diameter larger than 2.5 wavelengths and directivity larger than 17dBi. Since a SBFA roughly looks similar to a parabolic dish, a possible extension is to modify the large reflector of a SBFA towards a parabolic shape. Several different solutions have been described in the literature [2].

The simplest but not very efficient extension is the double rim SBFA. The latter includes a large reflector with two concentric rims. The inner rim is just a quarter wavelength high while the outer rim is one half wavelength high as shown in Fig 35.

The effect of two rims is barely appreciable. A maximum directivity increase of about 1dB can be expected when compared to a conventional single rim SBFA. As the directivity increases, the antenna becomes more sensitive to environmental conditions including the built in radome. The plots in Fig 36 show the difference between two double rim SBFAs for 13cm: one antenna without radome and the other with the aperture covered by a 1.6mm thick FR4 laminate:

As a conclusion, a double rim SBFA without radome provides about 0.5dB more directivity than a conventional SBFA at 2360MHz. Installing a 1.6mm thick radome from FR4 laminate, the

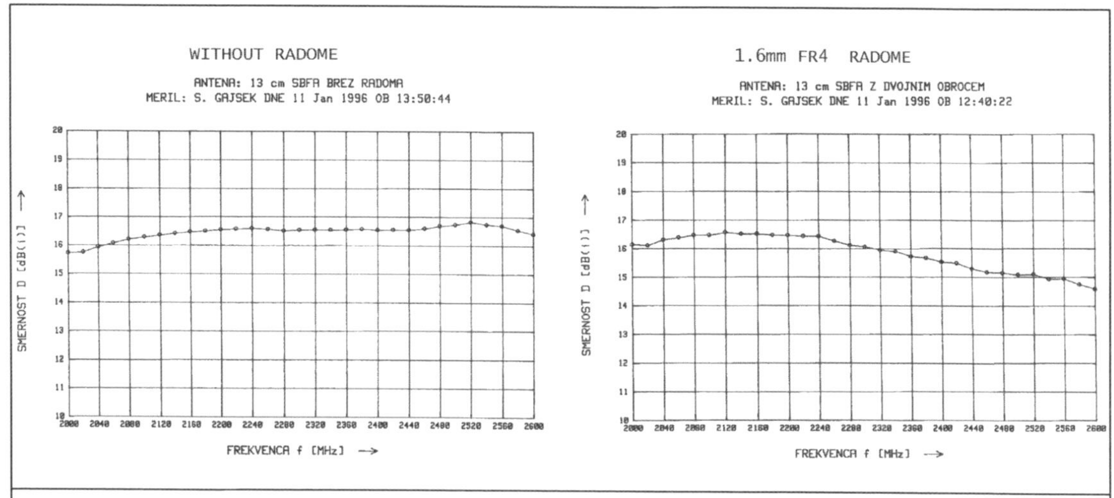

Fig 36: Radome effect on double rim SBFA directivity.

directivity drops by about 1dB and the final result is 0.5dB less directivity than a conventional SBFA. The double rim SBFA is therefore an academic curiosity with little practical value.

A side result of all measurements is that a radome made from FR4 laminate has a considerable effect on the SBFA performance already in the 13cm band. Therefore it is recommended to reproduce the described conventional SBFA with exactly the same materials, using 0.8mm thick FR4 or slightly thicker plexiglass for the radome.

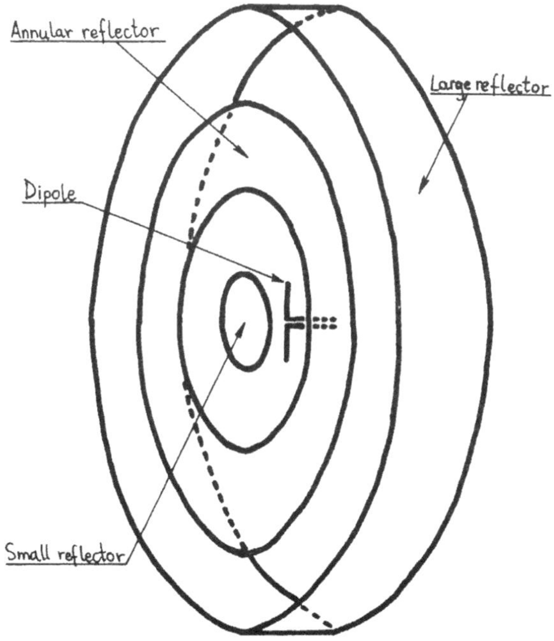

Fig 37: Archery target antenna structure.

Archery target antenna

As the SFBA cavity becomes larger, additional circular waveguide modes are excited. Rather than changing the shape of the large reflector, additional structures can be placed to control the amplitudes and phases of different modes. Thinking in terms of wave physics, a collimating structure in the form of Fresnel rings is required. The SBFA is already the first representative of such antennas, placing a small reflector in front of the feed dipole to control the lowest order Fresnel zone.

A further evolution of the above theory is a collimating structure including one small reflector disc and one annular reflector ring. Such a structure results in a rather efficient "archery target" antenna [4] as represented in Fig 37.

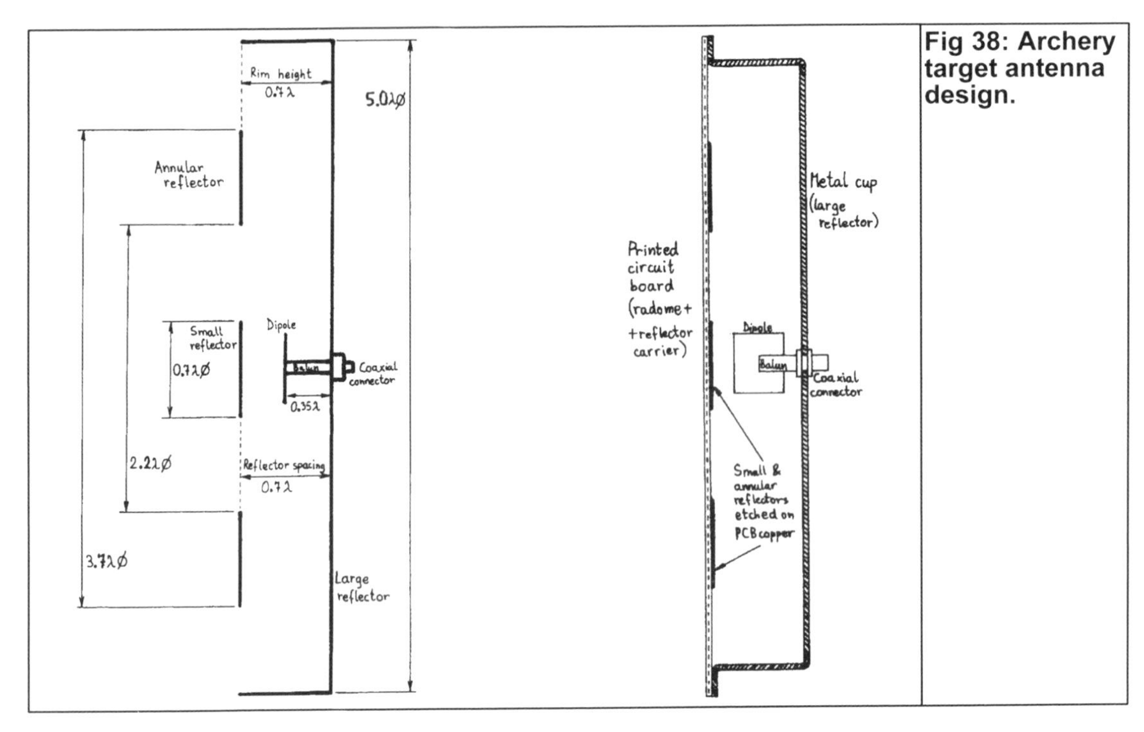

Fig 38: Archery target antenna design.

The "archery target" antenna described in this article achieves a directivity of 20.6dBi at an aperture efficiency of about 46%. The -3dB main lobe beamwidths are about 13.8 degrees in the E plane and 10.2 degrees in the H plane. This new antenna is simple to manufacture, since the supporting structure for the small and annular reflectors can perform as a radome at the same time.

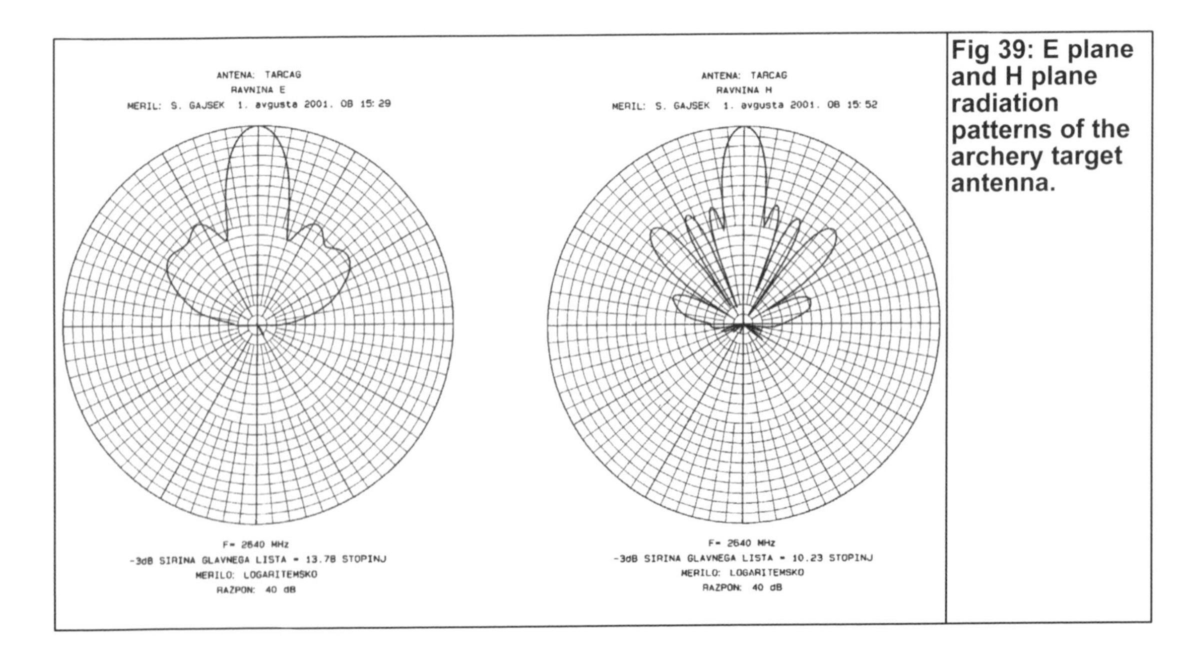

Fig 39: E plane and H plane radiation patterns of the archery target antenna.

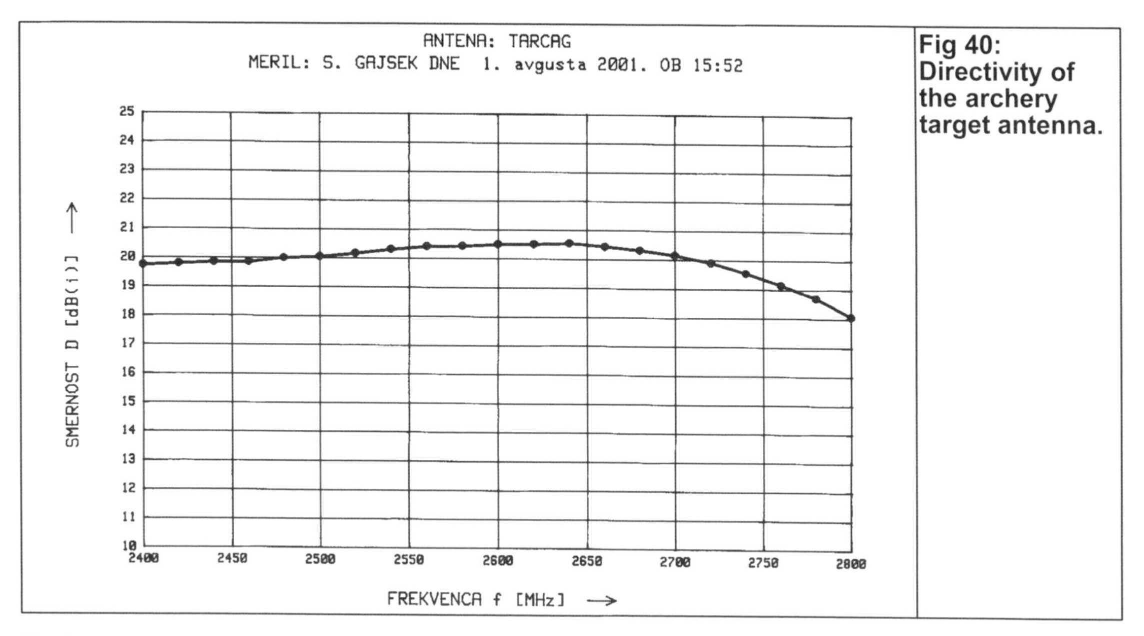

Fig 40: Directivity of the archery target antenna.

Probably the "archery target" antenna could be further optimised. Some computer simulations suggest that both a directivity of 22dBi and better aperture efficiency could be achieved although at a reduced bandwidth. Last but not least, the structure could be extended further to include several concentric annular reflectors.

The successful "archery target" antenna design presented in this article includes a large reflector with a diameter of about 5 wavelengths, much larger than in a typical SBFA. On the other hand, the small reflector has a diameter of 0.7 wavelengths and is comparable to the SBFA. The annular reflector extends from an inner diameter of 2.2 wavelengths to an outer diameter of 3.7 wavelengths. The reflector spacing and rim height are identical and equal to 0.7 wavelengths and Fig 38 is also somewhat larger than in a typical SBFA.

Fig 41: Dipole feed of the archery target antenna.

The prototype antenna has a large reflector diameter of 570mm, an annular reflector with an inner diameter of 252mm and an outer diameter of 420mm and a small reflector with a diameter of 80mm. The reflector spacing and rim height are set to 80mm. The small and annular reflectors are carried on a large dielectric plate: 0.8mm thick FR4 laminate with a dielectric constant of about 4.5. Although thin, this carrier plate has the effect of decreasing the optimum frequency by as much as 100MHz in the S band frequency range.

This prototype antenna achieved the best directivity of 20.6dBi at an operating frequency of 2640MHz. The measured E

Fig 42: Arcehry target antenna prototype.

plane and H plane radiation patterns shown in Fig 39.

The measured patterns in both planes at a number of different frequencies were used to compute the directivity as shown in Fig 40.

The first experiments with the "archery target" antenna were made with a simple thin wire half wave dipole feed. The dipole was positioned on the antenna axis of symmetry exactly halfway between the small and large reflectors just like in a SBFA. Since the thin-wire dipole had a poor impedance match to a 50Ω source even over a narrow frequency band due to the antenna cavity loading, several other feeds were experimented.

Reasonable impedance matching (-15dB return loss over a 10% bandwidth) was obtained with a single wide dipole feed built on a printed circuit board as shown in Fig 41.

While experimenting with different feeds, small (up to +/-0.2dBi) but repeatable variations of the antenna directivity were observed as well. In particular, the directivity decreased when the wide dipole printed circuit board was installed parallel to the reflector plates. On the other hand, the directivity improved when the wide dipole printed circuit board was installed perpendicular to the reflector plates.

The feed radiation pattern can therefore contribute to a more uniform illumination of both annular apertures of the "archery target" antenna. Effects of different feeds on other cavity antennas (cup dipoles and SBFAs) were not experimented yet. The complete archery target antenna prototype is shown in Fig 42.

Selection of the most suitable antenna

Although the whole family of microwave cavity antennas is very large, many of these antennas are not known to the wider public. Little if any serious articles have been published in the amateur radio literature. Therefore it was decided to write this article including the description of the most interesting microwave cavity antennas, their past experience, present performance

Fig 43: 23cm and 13cm SBFAs used for 1.2Mbps packet radio access.

and expected future developments. Fig 43 shows the 23cm & 13cm SBFAs used for 1.2Mbps packet radio access.

Unfortunately, most people select an antenna only according to its directivity or gain. WRONG! There are many more selection parameters and all of them need to be considered: width and shape of the main beam, side lobe levels and directions, frequency bandwidth, sensitivity to environmental conditions and weather effects, ease of manufacturing etc.

Microwave cavity antennas may not provide the maximum number of decibels for a given quantity of aluminium. This may explain why they are not so popular. On the other hand, microwave cavity antennas may be simple to manufacture, insensitive to manufacturing tolerances, have low side lobe levels, be reasonably broadband, easy to make weatherproof and insensitive to environmental conditions, quickly rejecting rain drops, snow and ice from accumulating on their radiating apertures.

While designing a radio link, the first consideration should be the antenna beamwidth according to the desired coverage. Antenna arrays should only be considered in a second place, when a single antenna is unable to provide the desired coverage. The most common mistake is to select the antenna with the largest number of decibels. Its beam may be too narrow, its large side lobes may pick interference and multipath and its deep nulls in the radiation pattern cause dropouts and pointing problems.

This article includes detailed descriptions of different cavity antennas: simple horn (7dBi & 90 degrees), cup dipole (12dBi & 50 degrees) and SBFA (16dBi & 30 degrees). All these designs are well tested and foolproof: it is just a matter of selecting the right design for a particular application. A SBFA is an excellent replacement and performs better than small (less than 1m diameter) parabolic dishes with poorly designed feeds at relatively low frequencies (below 2GHz).

On the other hand, a successful duplication of the "archery target" antenna (20.5dBi & 12 degrees) and its likely future developments requires some skill and appropriate test equipment. The intention of this article was to show that there is still development going on in the antenna field, providing some hints to serious antenna experimenters.

Antenna Array for the 6cm Band [5]

When one thinks about antennas for frequencies above a few GHz, the first types that come to mind are probably the parabolic reflector and the horn antennas. Both can be seen as "aperture" antennas, for which gain and directivity are related fundamentally to the ratio between the aperture area and λ^2.

The parabolic reflector antenna is conceptually simple, and can have large gain and high directivity. However, the construction of a high gain parabolic reflector can pose problems to an amateur because of the mechanical precision involved. Additionally, a good feed for a parabolic reflector isn't too easy to implement, especially if it has to be well matched to the dish. Due to losses in the feed, mismatch of the illumination and losses along the edge, the actual efficiency may be no more than 0.6 or so. Last, even if a shallow dish can be considered as an almost 2-D structure, the presence of the feed and its support transforms the parabolic reflector into a 3-D antenna.

The horn antenna, on the other hand, has the advantage that in most cases the construction is based only on plane surfaces. Also, a horn has wide bandwidth and its construction is tolerant to dimensional errors. However, the horn suffers from the same basic drawback as the dish antenna: both are fundamentally 3-D antennas, as the length of a high-gain horn is comparable to its lateral dimensions.
An array of dipoles can be an interesting alternative to both the parabolic reflector and the horn antennas for the 5.7GHz band, especially if the longitudinal dimensional of the antenna must be minimum. Our intention here is to show that it is relatively easy to construct a broadside antenna array for moderately high gains, say, between 15 and 25dBi. A dipole array is a low profile antenna that can be lightweight, cheap and simple to construct, and yet have good performance. The reflector for an antenna array is plane, not curved, so it is much simpler to construct than, for example, a parabolic one. If more gain is desired, it is easy to scale up the linear array into a bidimensional array with larger gain. Although the described antenna array in this article was designed for 5.2GHz, it is relatively easy to adapt the design for the 5.7GHz amateurs' band. We hope this article will encourage other people to experiment with antenna arrays.

An antenna array is basically a collection of elementary antennas working cooperatively to reinforce the transmitted field (or increase the sensitivity, if the it is a receiving antenna) at some directions and weaken at others. The elementary antennas are fed by a signal distribution system. The feeding system is very important because it determines the antenna directivity and gain, the direction of maximum radiation or maximum sensitivity, the input impedance, etc. Certain radars, like those used with Patriot missiles, use sophisticated phased arrays antennas. In such antennas, variable phasing circuits in the feed system are used to electronically scan the space under observation. Traditional radars, in contrast, scan the space mechanically, by rotating a parabolic antenna.

In the antenna array design to be described (Fig 44), the goal was maximum simplicity, not sophistication or ultimate performance. The construction was based on materials widely available. The radiators are dipoles with a nominal length of one wavelength. Each radiator is fed with a signal of same amplitude and in phase. That is, the phase is 0° for all dipoles. The array is of "broadside" variety and the direction of maximum propagation is perpendicular to the plane of dipoles. A flat reflector behind the dipoles increases the gain at the same time the

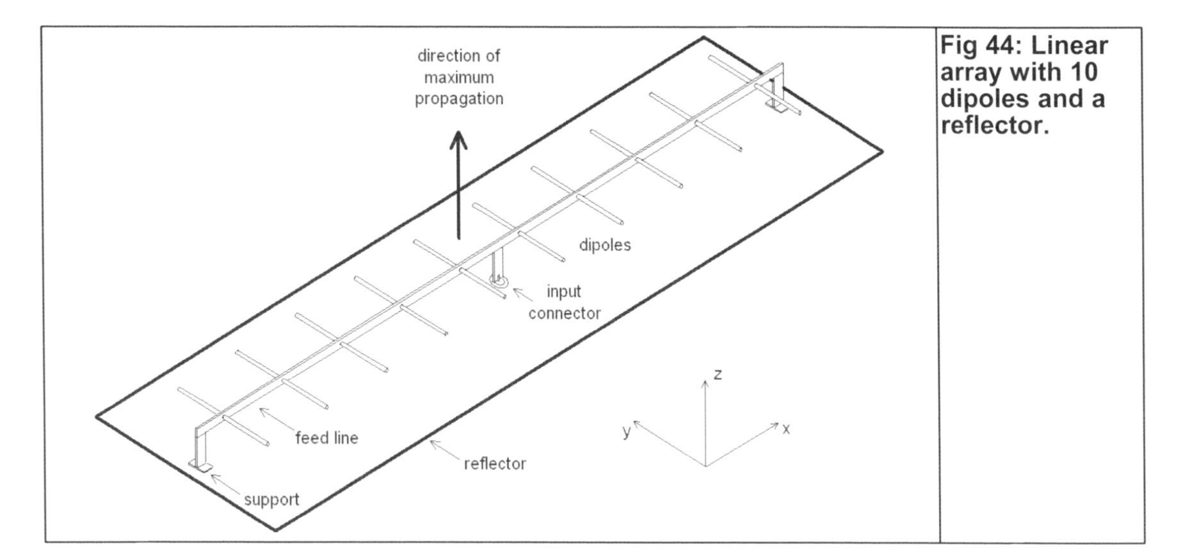

Fig 44: Linear array with 10 dipoles and a reflector.

backward propagation is virtually eliminated.

The feed line is simply an FR-4 strip. It does the electrical work of sending the signal to the various radiators and is used also as a mechanical support for the dipoles. The reflector is an FR-4 panel with the copper side toward the dipoles. An SMA connector is attached to the center of the FR-4 panel and serves as the input for the antenna. We will suppose it is a transmitting antenna but, because of the reciprocity property, most of the conclusions will also apply when the antenna is receiving.

Although the antenna array appears deceptively simple in principle, several practical and theoretical problems must be solved before we have an operational antenna. Because "God is in the details", let us briefly examine the theoretical principles of an antenna array.

Theory of antenna array

The Fig 45 shows the linear array, as seen with the axis y normal to the page. The elements of the N-array is represented by the points 0, 1, 2, ... N-2, N-1. It is supposed that the current in

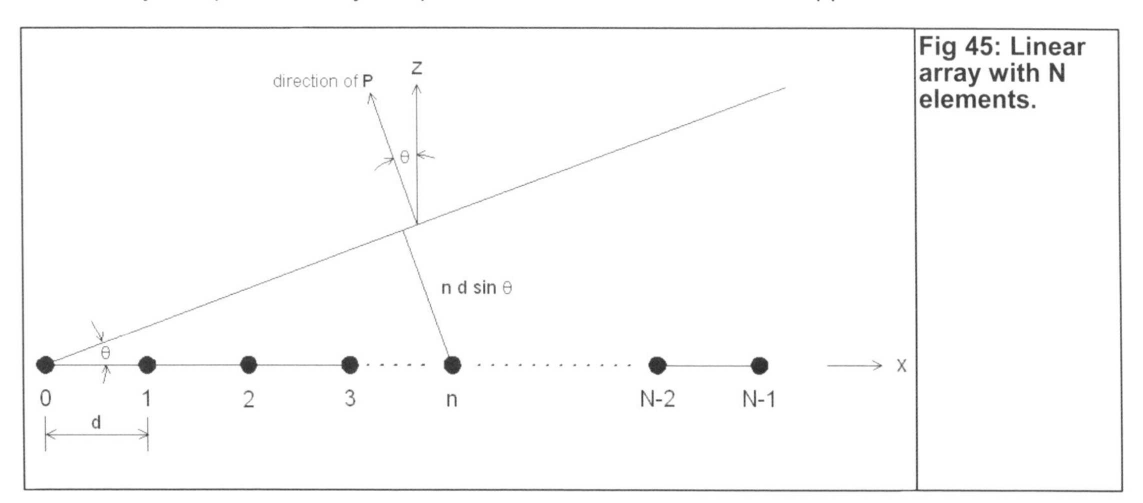

Fig 45: Linear array with N elements.

each element of the array have same amplitude and phase, so the z is evidently the direction of maximum radiation.

A point P, very far from the array, will receive in the general case a sum of waves with different phases (but practically equal amplitudes) since the distance between P and each element of the array will vary. If r is the distance from P to the element 0, the distance from P to the element 1 is (r + dsinθ), the distance from P to the element 2 is (r + 2dsinθ), and so on. The field intensity at P is given by:

$$E = kF(\theta)\sum_{n=0}^{N-1} e^{-j\frac{2\pi n d\sin\theta}{\lambda}} \qquad (1)$$

where

- k is a complex constant that accounts to the path loss and the transmitting power
- F() is the directivity function of the array elements
- $\sum_{n=0}^{N-1} e^{-j\frac{2\pi n d\sin\theta}{\lambda}}$ is the directivity function of the array.

Note that θ is given in radians in the formula above. Note also that k is independent of θ if the propagation mean is uniform. The directivity (and the gain) of any array depends on the intrinsic directivity F(θ) of each array element and on the directivity function of the array. If the array elements are isotropic, that is F(θ) = 1, an array can be very directive because of the directivity function of the array.

The behaviour of the directivity function can be studied from the diagram of Fig 46, which shows that the directivity function can be seen as the resultant of a summation of N complex phasors.

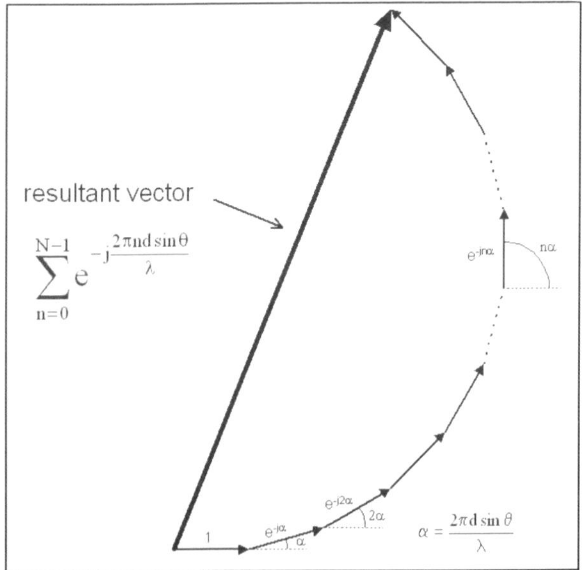

Fig 46: Geometric interpretation of the array directivity function.

The sequence of phasors is a geometric progression so the summation of the N complex phasors can be calculated easily:

$$\sum_{n=0}^{N-1} e^{-j\frac{2\pi n d\sin\theta}{\lambda}} = e^{-j(N-1)\pi d\sin\theta}\,\frac{\sin\left(\frac{N\pi d\sin\theta}{\lambda}\right)}{\sin\left(\frac{\pi d\sin\theta}{\lambda}\right)} \qquad (2)$$

Usually we are interested only on the absolute value of the directivity function:

$$A(\theta) = \left|\sum_{n=0}^{N-1} e^{-j\frac{2\pi n d\sin\theta}{\lambda}}\right| = \left|\frac{\sin\left(\frac{N\pi d\sin\theta}{\lambda}\right)}{\sin\left(\frac{\pi d\sin\theta}{\lambda}\right)}\right| \qquad (4)$$

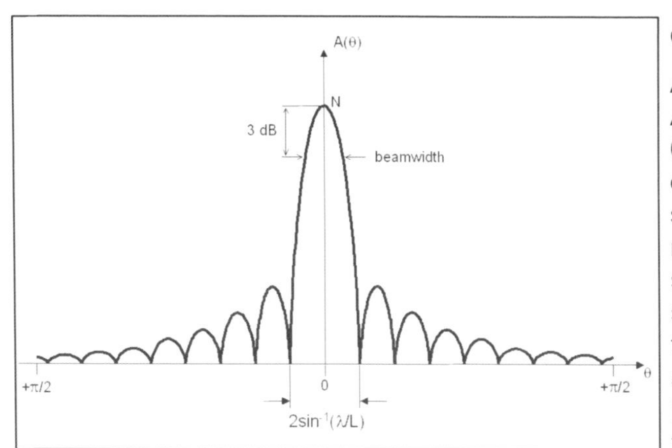

Fig 47: Absolute value of the directivity function as a function of θ.

Gain an directivity

A(θ) has a maximum at θ = 0, where A(θ) = N. This isn't a surprise since θ = 0 is the direction for which all the elements of the array send waves with same phase. In the direction of maximum gain, the far field is N times stronger than the field produced by a single element. It is easy to think that the power gain of the array compared to an antenna with a single element would be N^2, since the power is proportional to the square of the field intensity. In fact, the power gain is only N because the input power for the total array is N times the input power for each element.

Fig 47 shows how A(θ) varies with θ. It is enough to represent A(θ) for θ between –π/2 and +π/2 because A(θ) is periodic with a period equal to π. Each time the resultant vector of Fig 46 makes a full turn, A(θ) reaches a null.

The main lobe is centred at θ = 0 and its width, measured as the distance between nulls, is given by:

$$\Delta\theta_{NULLS} = 2\sin^{-1}\left(\frac{\lambda}{Nd}\right) \qquad (5)$$

The secondary lobes, also called sidelobes, have half the width of the main lobe. Note that L = Nd is the approximately equal to the physical length of the array, so the main lobe width is given by:

$$\Delta\theta_{NULLS} = 2\sin^{-1}\left(\lambda/L\right) \qquad (6)$$

The result above shows that the width of the main lobe varies with the inverse of the array length given in wavelengths.

The usual way to evaluate the directivity of an antenna is through the so-called beamwidth, defined as the angular aperture for –3dB. The beamwidth is smaller than the angular aperture given by the first nulls of the radiation pattern. There is no closed formula to calculate the beamwidth, but a good approximation can be found if one notes that in the equation (4) the argument of the sine at denominator is small for in the main lobe interval, so the sine can be approximated by its argument:

$$A(\theta) \cong \left|\frac{\sin\left(\frac{N\pi d\sin\theta}{\lambda}\right)}{\frac{\pi d\sin\theta}{\lambda}}\right| = N\left|\frac{\sin\left(\frac{N\pi d\sin\theta}{\lambda}\right)}{\frac{N\pi d\sin\theta}{\lambda}}\right| = N\left|\frac{\sin\left(\frac{\pi L\sin\theta}{\lambda}\right)}{\frac{\pi L\sin\theta}{\lambda}}\right| \qquad (7)$$

The equation (7) shows that A(θ) is a (sin x)/x type function, that has a peak at x = 0 and decays 3dB at x = ±1.391 (radians). In terms of beamwidth, this corresponds to:

$$\Delta\theta_{-3dB} = 50.7\left(\lambda/L\right) \quad \text{(degrees)}$$

Table 1 shows how the lobe and the beam width vary for several array lengths:

Table 1: Lobe and beamwidth with varying array length.

L/λ	Lobe width (degrees)	Beamwidth (degrees)
2	60	25.4
3	38.9	16.9
5	23.1	10.1
10	11.4	5.1
20	5.6	2.5

Grating lobes

In certain conditions, an antenna array can produce side lobes with same amplitude of the main lobe. In general, these grating lobes are undesirable in an antenna. A grating lobe occurs if for some θ, dsin(θ) (Fig 45) is equal to a integer multiple of one wavelength. To guarantee that grating lobes will not occur, it is necessary that the distance d between consecutive array elements be less than a wavelength. A separation commonly used between dipoles is which is very convenient value, as we will see when discussing a feed line based on a strip line.

The reflector

The flat reflector placed behind the dipoles has two main effects. One is suppressing the backward radiation; the other is increasing the gain of the array. The gain increases because the energy that is radiated backward is reflected back by the reflector and so it adds to the direct radiation and finally goes to the "right" hemisphere. It is known that an electromagnetic wave suffers a phase reversion after a reflection on a conductor, so it is intuitive that a good position to the reflector is when its distance to the dipoles is equal to λ/4. In this way, the backward wave travels a distance equal to 2 x λ/4 = λ/2 before joining the direct wave, as shown in Fig 48. The λ/2 total path for the reflection introduces a phase variation of 180° but as a metallic reflector introduces a further 180° phase variation, the reflected and the direct radiate in phase. Nonetheless, it can be a little surprise that the gain increases slightly if the distance between the dipole and the reflector is made shorter than λ/4. On the other hand, the antenna bandwidth decreases and the sensitivity for losses increases for shorter distances, so λ/8 is the minimum recommended separation between the dipole and the reflector.

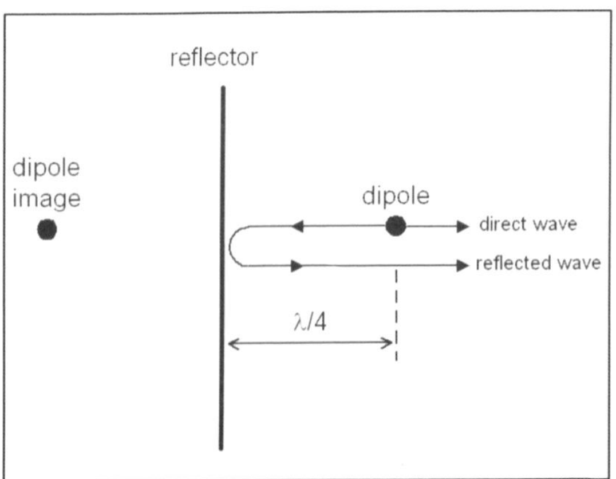

Fig 48: Wave reflection by the reflector.

As an observation, the flat reflector is replaced by an electromagnetic image of the dipole in the mathematical analysis called "method of images". The current in the image dipole is inverted relative to the real dipole, so the tangential electric field and the normal magnetic field are both zero on the surface of the reflector, which is usually considered as a perfect conductor.

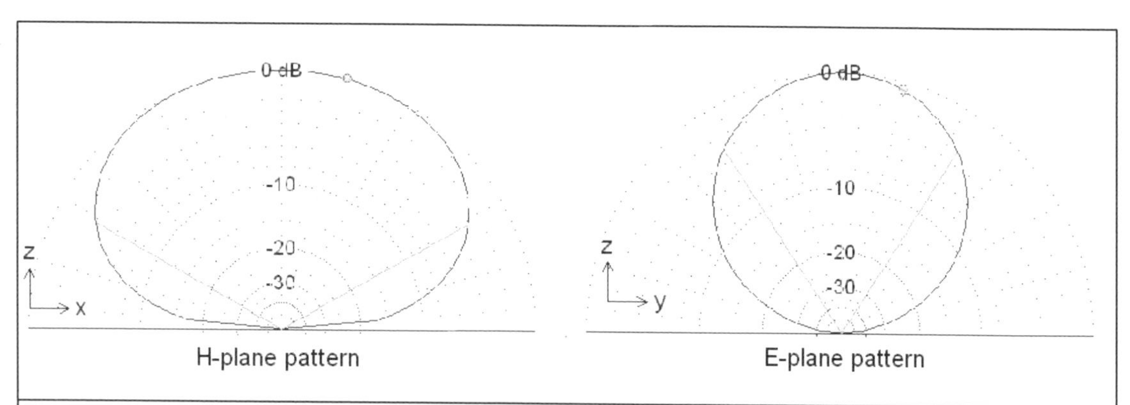

Fig 49: Radiation patterns for a full-wave dipole over a conductor plane.

Coupling between the dipoles

The coupling between the dipoles of an array can alter the radiation pattern and the input impedance of each dipole. However, the coupling isn't too strong if the distance between successive dipoles is around $\lambda/2$, so it is usual to ignore the coupling except in the cases of much sophisticated projects of antenna arrays.

Designing the dipoles

In many types of antennas based on the dipole, the length of the dipole(s) is about a half wavelength. A half-wave dipole is resonant and its input impedance is about 73Ω, a value that is easily matched to a 50Ω transmission line. However, in an antenna array the dipoles are effectively connected in parallel by the feed line, so the input impedance can be inconveniently low, especially if the number of half-wave dipoles in the array is high. So, in large arrays, the full-wave dipoles are preferred over half-wave dipoles because the input impedance of a full-wave dipole is much higher than that of a half-wave dipole. The use of full-wave dipoles eases considerably the matching to the transmission line.

A full-wave dipole doesn't have a well-defined input impedance like the 73Ω of a half-wave dipole. The input impedance of a full-wave dipole can be hundreds or thousands of ohms, depending on the ratio of the diameter d to the length l. Theoretically, as d/l tends to zero, the input impedance tends to infinity. In practical terms, a thin full-wave dipole can have high input impedance (>1000Ω) and a thick dipole has moderately low input impedance (hundreds of Ω). At microwave frequencies, the dipoles are generally thick because l is relatively small. An advantage of the thick dipole is its greater bandwidth, since the input impedance varies less with frequency than that of a thin dipole. It should be noted, too, that a thick full-wave dipole is resonant at a frequency significantly lower than that corresponding to one wavelength. In other words, a "full-wave" dipole for microwave can be considerably shorter than one wavelength.

In the project, the dipoles were made of 2.26mm diameter solid copper wires commonly used in electrical installations. The length that resonates in 5.2GHz was determined by trial-and-error using the EZNEC simulation software. The free-space wavelength for 5.2GHz is 57.7mm but it was found that the 2.26mm diameter full-wave dipole resonate at 5.2GHz when the length is 34mm, just 59% of one wavelength.

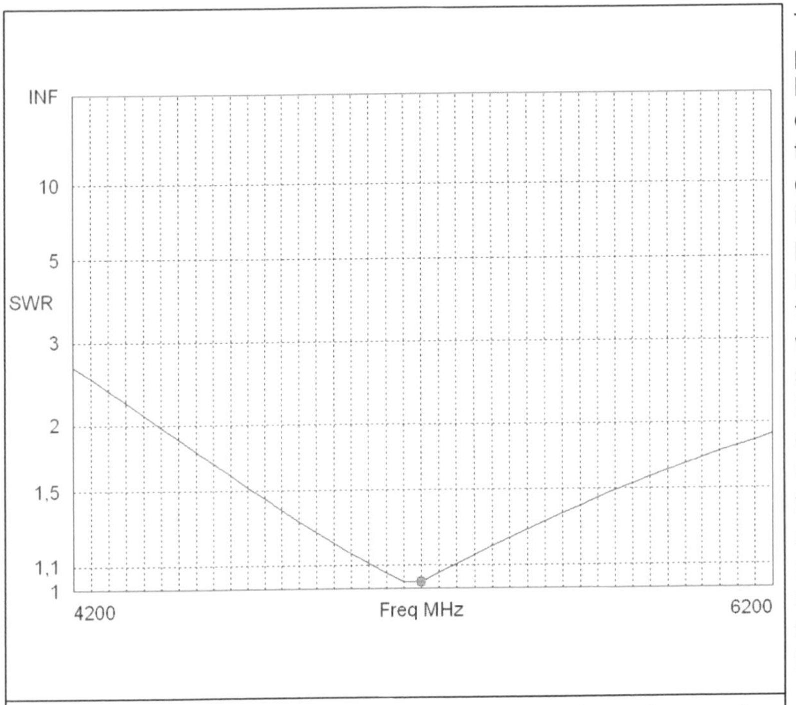

Fig 50: SWR of a single dipole for a reference impedance of 300Ω.

The distance from the dipole to the reflector plane has a significant influence on the input impedance, the gain and the directivity of the dipole. Besides, the input impedance of the dipole gets higher as it is moved toward the reflector. By the way, a half-wave dipole has an opposite behaviour, that is, the input impedance decreases as the dipole comes close to the reflector. As a practical rule, the distance between the dipole and the reflector should be about a quarter wavelength. In the project, it was used separation of 15mm, which results in an input impedance nearly equal to 300Ω.

Fig 49 shows the radiation patterns for a single dipole at a distance of $\lambda/4$ from an infinite conductor plane. The beamwidth for the H plane (plane xz of Fig 44) is a wide 122°. The radiation pattern is very broad, almost omni-directional. The radiation pattern would be a perfect circle if the reflector didn't exist. Nevertheless, an array with 10 dipoles is very directional in the H-plane because of the array directivity function, as shown by equation (1).

The beamwidth in E-plane (yz plane of Fig 44) is 67°, which is narrower than the beamwidth in H plane. However, it should be noted that the E directivity of the linear array doesn't benefit from the array directivity function since the dipoles are lined up along the x axis.

The 2.26mm diameter 34mm long dipole has a gain of 7.65dBi and an input impedance of 307Ω at 5.2GHz, as predicted by EZNEC. Fig 50 shows how the SWR varies with frequency for a reference impedance equal to 300Ω. The SWR varies slowly and is less than 2 for frequencies between 4.5 and 6GHz. This broadband behavior is basically the result of the thickness of the dipole, which has d/l equal to 6.6%. In contrast, a dipole for HF bands can have d/l as low as 0.002%.

Designing the feed line

The feed network is probably the most critical component of an antenna array. In a broadside array, the feed network must feed all dipoles in phase. The feed network is made of multiple transmission line sections. Each section must have the correct length because in a transmission line the phase is a function of the line length.

In many arrays, the dipoles are separated by half wavelength, as shown in Fig 51. Note the

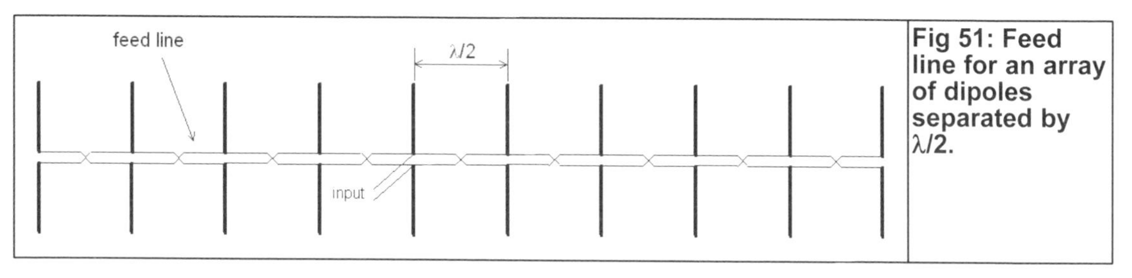

Fig 51: Feed line for an array of dipoles separated by λ/2.

transpositions in the feed line are used to compensate the 180° phase introduced by the propagation in a λ/2 long transmission line sections. Of course, this method of feeding works only if the phase velocity in the feed line is the same as in free space, for example, when the feed line uses air as dielectric, like the cases where bare wires in "X" are used as feed line.

It is important to note a property of a transmission line of length λ/2 and characteristic impedance Z_0. If the termination load is Z_L, then the input impedance of the line is equal to Z_L, no matter the values of Z_0 or Z_L. This property holds for any length multiple of λ/2, like , 3λ/2, 2λ, etc. Even if the feed line SWR isn't equal to one, the phasing of each section will be always 0° or 180°. The input of the array can be the terminals of any dipole but in general it is better to place the input at the center of the array. The input impedance of the array is Z_d/N, where Z_d is the input impedance of each dipole, and N is the number of dipoles.

In the project, a strip of double-sided FR-4 board with a thickness of 1.5mm was used as a feed line. The strip also serves as a mechanical support for the dipoles, which are soldered directly to the strip line {see footnote 1} as shown in Fig 52. The strip can be easily cut with a saw from a virgin FR-4 board. The exact width isn't critical and something like 3mm is adequate.

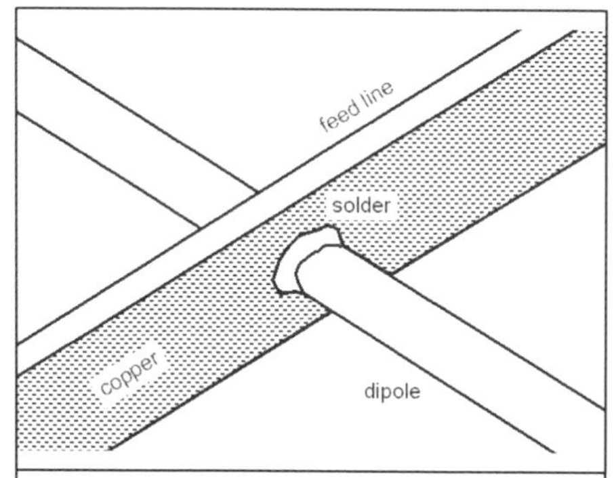

Fig 52: Detail of the dipole soldered to the feed line.

To successfully implement a broadside array, it is necessary that:

- The dipoles work in phase
- The physical distance between them don't depart too much from λ/2

Fortunately, the FR-4 material has the "right" dielectric constant to satisfy both requirements. The (relative) dielectric constant, ε_r, of the FR-4 dielectric is 4.65 but as in the strip line, part of electric field is in the air, ε_{eff}, the effective dielectric constant, is about 3.7. The velocity of propagation in the strip feed line is given by $c/\sqrt{\varepsilon_{eff}}$, where c is the free-space velocity. Using a value of ε_{eff}, = 3.7, the calculated velocity in the strip feed line is about 52% of the free-space velocity that means the guided wavelength, λ_g, is equal to 0.52λwhere is the wavelength in free space. Note that λ_g is the length that changes the phase of the guided wave by 360°. So if the physical separation between dipoles is made equal to λ_g, that is almost equal to λ/2, the requirements for correct feeding are of satisfied, even without using transposition.

Measuring the velocity of propagation in the FR-4 strip line

The distance between dipoles is probably the most critical dimension of the array. As described before, for a strip line feed based on FR-4 substrate, the distance between dipoles must be exactly one λ_g, the guided wavelength. The method used for determination of λ_g was based on

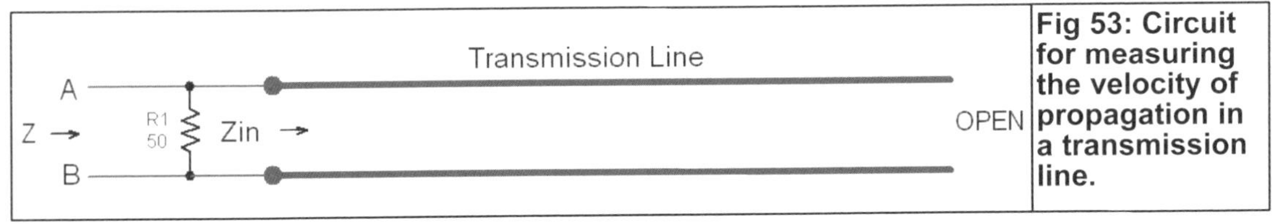

Fig 53: Circuit for measuring the velocity of propagation in a transmission line.

measurement of the return load of an open-loaded transmission line with its input in parallel with a resistance of 50Ω, as shown in Fig 53.

Suppose first that the transmission line is loss-free. The input impedance Zin of an open-loaded transmission line varies from zero to infinity, depending on the frequency. If the transmission line length L is a multiple of $\lambda_g/2$, Zin is infinite, but if L is an odd multiple of $\lambda_g/4$, then Zin is zero. It is easy to see that Z, which is the combination of Zin in parallel with R1, varies from zero to 50Ω, depending on the frequency. The return loss (S11) measured in dB at the point A and B will vary, as a function of frequency, between zero and minus infinity. Theoretically, the RL is minus infinity when the input impedance of the transmission line is infinite and will be zero when the input impedance of the line is zero.

On the other hand, if the transmission line is lossy, RL will oscillate from something below zero and something above minus infinity, as show in Fig 54 for a FR-4 strip line with width of 3mm and length of 355mm. The graph is from a Sonnet simulation; high-frequency electromagnetic software which can be downloaded free from http://www.sonnetsoftware.com/.

As frequency increases, the line losses also increase but the amplitude of the oscillation of RL decreases. Furthermore, as $f \rightarrow \infty$, $Z_{in} \rightarrow Z_0$ (Z_0 is the characteristic impedance of the transmission line) and RL $\rightarrow -9.5$dB ($Z \rightarrow Z_0/2$). Note that the positive peaks correspond to odd multiples of $\lambda_g/4$ and the negative peaks to multiples of $\lambda_g/2$.

For measurement of the return loss, two 100Ω SMD resistors were soldered to the back of a female connector (it can be BNC, TNC or SMA). The resistors were soldered directly from the central pin to the outer conductor. This is important to keep the parasitic inductances to a

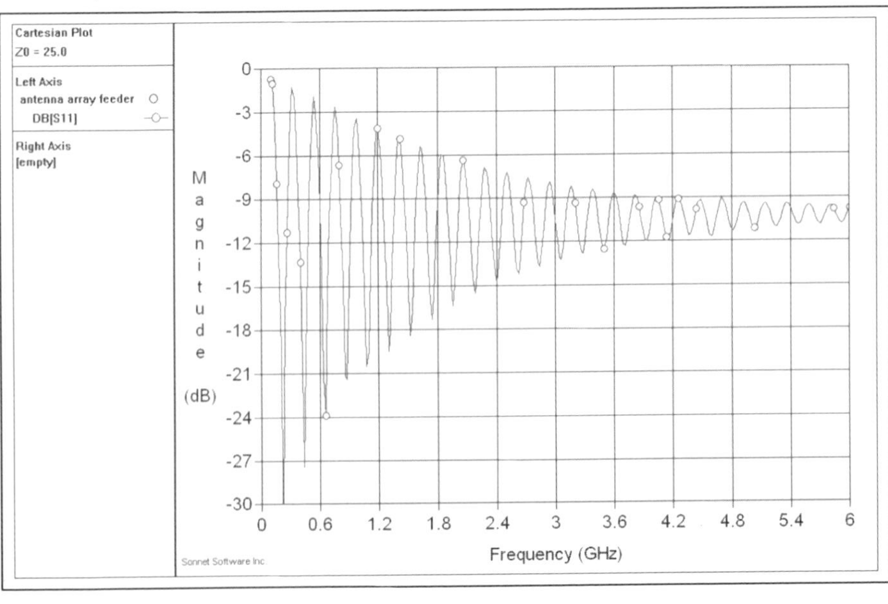

Fig 54: Return loss of a 3mm wide and 355mm long FR-4 strip line.

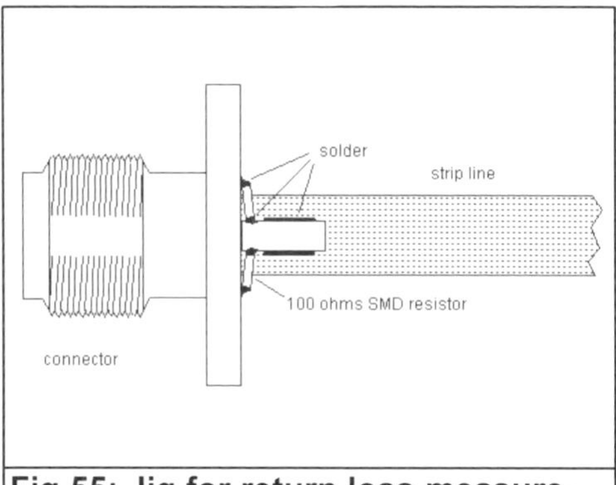

Fig 55: Jig for return loss measurement.

minimum. The end of the strip line was soldered to the central pin and to the ground as shown in Fig 55.

Fig 56 shows the result of the measurement of the return loss. A spectrum analyser with a tracking generator and an RF bridge was used. Note that the marker at 5.2GHz is on the 22nd positive peak. This means that the length of the strip line corresponds to 43 times $\lambda_g/4$ at 5.2GHz. The length of the strip line is 355mm, so λ_g is equal to 355 x 4/43 = 33.0mm, that is also the distance to be used between dipoles. The velocity of propagation calculated from λ_g is 171,700km/s, that is 57.2% of the velocity of propagation in free space.

From the curve of RL simulated by Sonnet (see again Fig 54), it can be calculated that g is equal to 30.2mm and that the velocity of propagation in the FR-4 strip line is equal to 52.4% of the free-space velocity. Comparing the velocity propagation simulated with the measured value there is discrepancy of nearly 10%. The velocity in the real strip line is greater than the simulated value. This discrepancy can be explained for the dielectric layer in the real line is truncated, whereas it is not in the simulation. To understand this point better, see Fig 57 that shows the section views of the strip line, real and simulated.

The real strip line (A) is a balanced structure with a width of 3mm and a thickness of 1.5mm. The characteristic impedance Z_0 is about 50Ω. This balanced line can be analysed from an equivalent unbalanced line (B) with half the dielectric thickness and half the characteristic impedance.

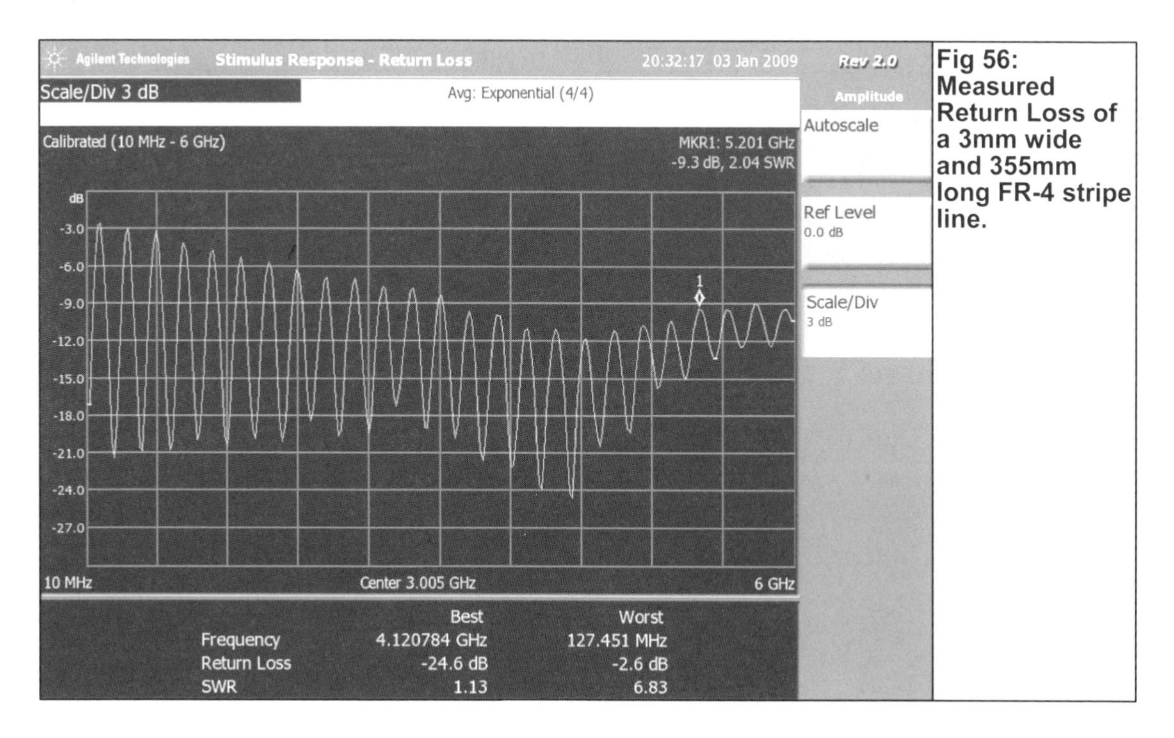

Fig 56: Measured Return Loss of a 3mm wide and 355mm long FR-4 stripe line.

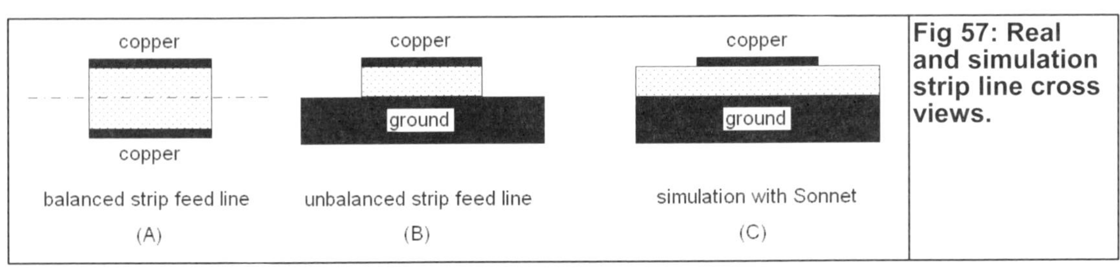

Fig 57: Real and simulation strip line cross views.

Sadly, Sonnet Lite doesn't allow truncated dielectric layer, so the simulation was done for the continuous dielectric layer (C). This simulated microstrip has a greater fringe capacitance than the real strip line, because in the simulated microstrip, the most part of fringe field lines are in the dielectric, which has greater dielectric constant than the air.

Feed line losses

Losses in FR-4 increase rapidly with frequencies, so one could be sceptical if FR-4 is appropriate as a substrate for a feed line operating in 5.2GHz. We will show that the losses aren't excessive.

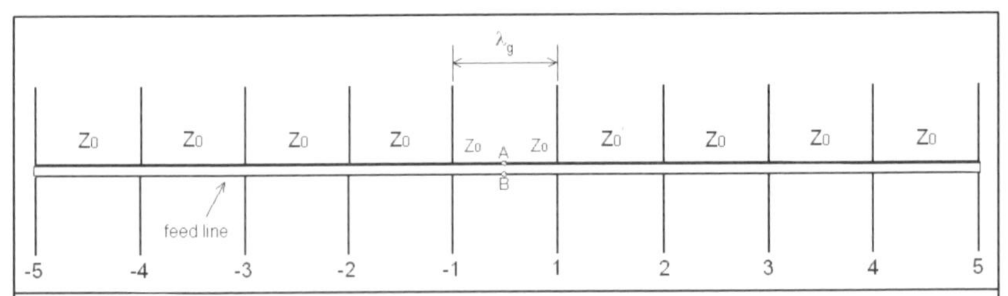

Fig 58: Feed line with uniform impedance.

In the prototype a feed line with uniform characteristic impedance was used (Fig 58). The central points A and B are the input for the array. Each line section between consecutive dipoles has the same

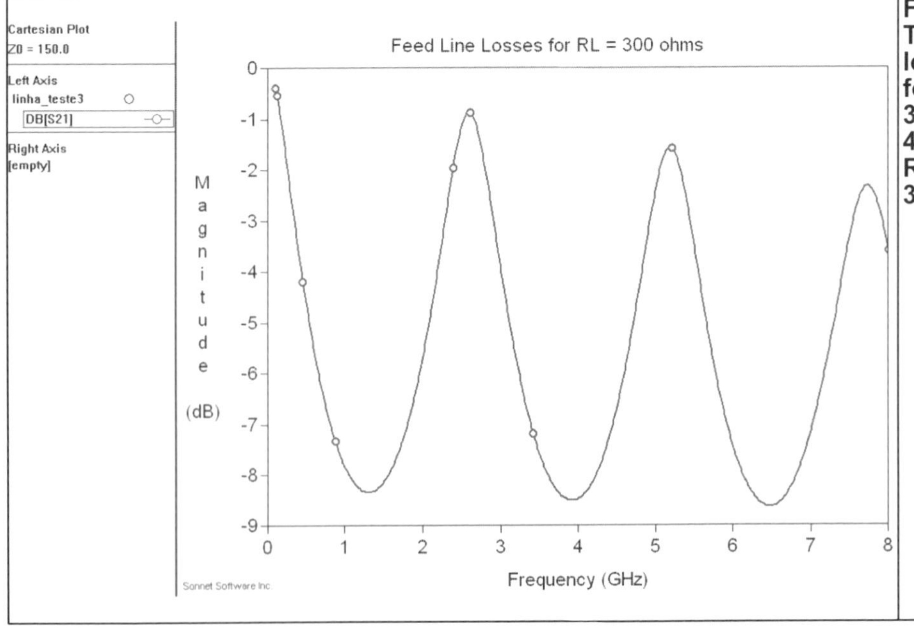

Fig 59: Transmission loss simulated for a 3mm wide, 30mm long, FR-4 strip line with RL equal to 300Ω.

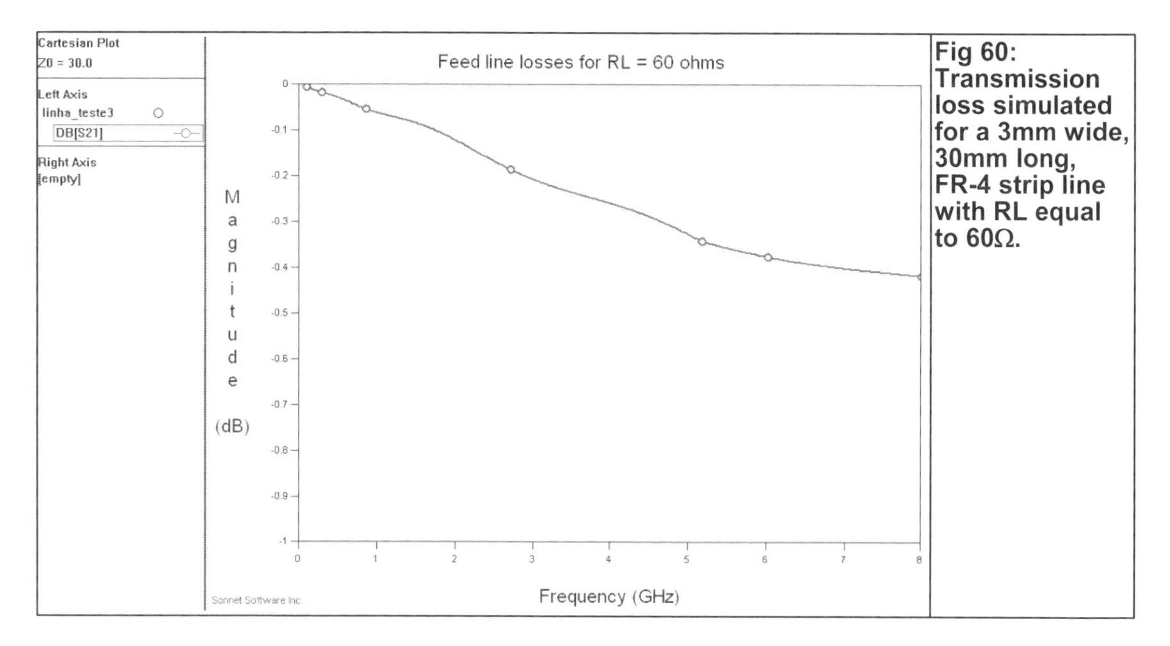

Fig 60: Transmission loss simulated for a 3mm wide, 30mm long, FR-4 strip line with RL equal to 60Ω.

characteristic impedance Z_0. However, the load impedance for each section varies. For example, the last right line section is loaded by just the dipole 5, but the second last right line section is loaded by the dipole 4 in parallel with the impedance of dipole 5 reflected by the last right line section. As each line section has length equal to λ_g or $\lambda_g/2$, the load at points 1, 2, 3, 4 and 5 are RL/5, RL/4, RL/3, RL/2 and RL, respectively, where RL is the input impedance of a single dipole (which is 300Ω in the project, as seen before). The impedance seen by the generator at input AB is RL/10, or 30Ω.

The losses in a FR-4 strip line, 3mm wide and with length equal to λ_g for 5.2GHz, were simulated by the Sonnet software. The line was connected between port 1 and port 2 of the simulator. The input signal was injected at port 1 and port 2 was terminated with a resistive load with a variable value.

Fig 59 shows the simulated transmission loss (S21) as a function of frequency for a 300Ω load. The minimum loss occurs at the frequencies where the line length is equal to multiples of $\lambda_g/2$, that is, for 2.6, 5.2, and 7.8GHz. For these particular frequencies, the input impedance is the same as the termination, as a half-wave line reflects the load impedance to the input. For other frequencies, the transmission loss increases because of the mismatch between the generator and the line input.

Table 2: Data of relative power received.

R_{LOAD} (Ω)	Segment length	Loss at 5.2GHz (dB)	(%)
300	λ_g	1.57	30.3
150	λ_g	0.93	19.3
100	λ_g	0.74	15.7
75	λ_g	0.68	14.5
60	$\lambda_g/2$	0.34	7.5

Fig 60 shows the transmission loss when RL is made equal to 60Ω. This load is nearly perfectly matched to the line. In this situation, the transmission loss increases gradually with the frequencies and the ripple almost disappears.

Table 2 shows the transmission loss for the various section of the feed line, when terminated by 300, 150, 100, 75 and 60Ω. The line section with the greatest loss is the one feeding dipole 5

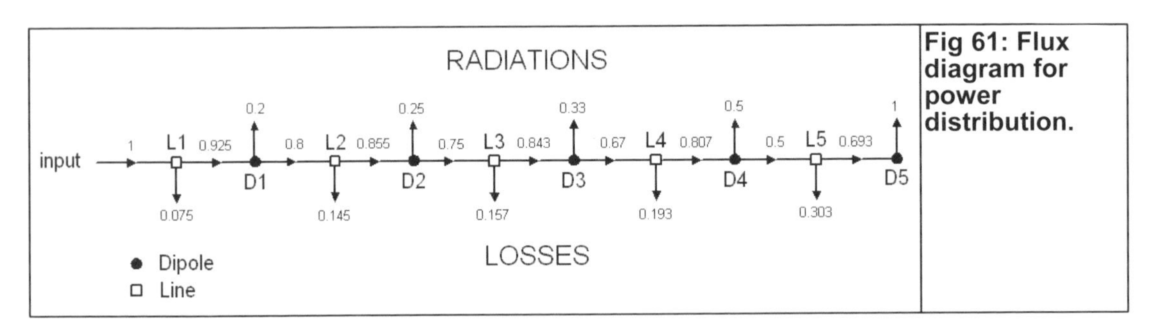

Fig 61: Flux diagram for power distribution.

that is the most mismatched. The half section between the input and dipole 1 correspond to the lowest loss because of the good match and, principally, because it is the shortest one.

With the data from the Table 2 it is possible to calculate the relative power received by each dipole and so to estimate the total loss of the feed system. Fig 61 shows how the power is distributed in the antenna array. For example, the line section L1 between the input and the dipole D1 looses 7.5% of the power but delivers 92.5% to the terminals of D1. One fifth (20%) of the power that reaches the D1 terminals are radiated by dipole D1 and four fifth (80%) is delivered by L2 to the next dipoles, and so on.

The power radiated by each dipole can be calculated by:

- $P_{D1} = 0.925 \times 0.2 = 0.185\ P_{in}$
- $P_{D2} = 0.925 \times 0.8 \times 0.855 \times 0.25 = 0.158\ P_{in}$
- $P_{D1} = 0.925 \times 0.8 \times 0.855 \times 0.75 \times 0.843 \times 0.33 = 0.132\ P_{in}$
- $P_{D1} = 0.925 \times 0.8 \times 0.855 \times 0.75 \times 0.843 \times 0.67 \times 0.807 \times 0.5 = 0.108\ P_{in}$
- $P_{D1} = 0.925 \times 0.8 \times 0.855 \times 0.75 \times 0.843 \times 0.67 \times 0.807 \times 0.5 \times 0.693 = 0.075\ P_{in}$
- $P_{TOTAL} = 0.658\ P_{in}$

The total lost power in the feed line is 0.342 P_{in}, that represents a total loss of 1.8dB. As a reference, a parabolic antenna with an efficiency of 50% wastes 3dB of the power, so a loss of 1.8dB in the feed line based on a FR-4 substrate appears to be reasonable.

Optimising the feed line

The loss in the feed line can be reduced if the characteristic impedance Z_{0n} of the each line section is matched to the resistance seen at each feed point. Referring now to Fig 62 and from the previous discussion, the characteristic impedance of each line section for a matched condition is:

- $Z_{05} = 300\Omega$
- $Z_{04} = 150\Omega$
- $Z_{05} = 100\Omega$
- $Z_{05} = 75W$
- $Z_{05} = 60W$

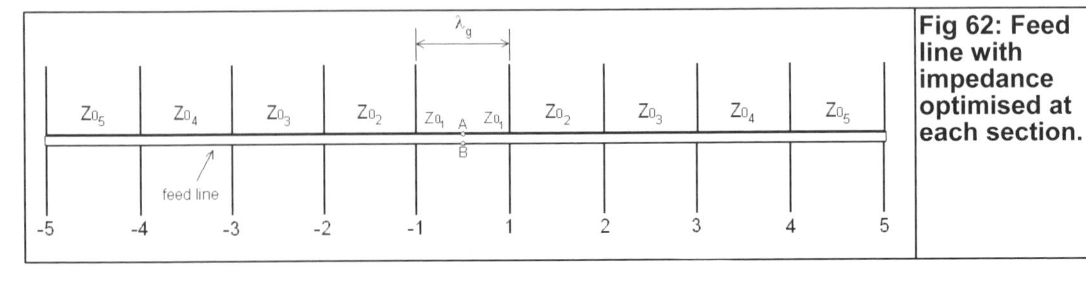

Fig 62: Feed line with impedance optimised at each section.

Table 3: Revised data of relative power received.

R_{LOAD} (Ω)	Line length	Loss at 5.2GHz (dB)	(%)	Line width (mm)
300	λ_g	0.60	12.9	0.025
150	λ_g	0.64	13.7	0.75
100	λ_g	0.65	13.9	1.50
75	λ_g	0.66	14.1	2.25
60	$\lambda_g/2$	0.34	7.5	3.00

Table 3 shows that the loss is reduced by a little more than ½ dB for the g long sections. The $\lambda_g/2$ is the same as the non-optimised feed line. And note how narrow some lines are now, especially that for 300Ω.

Calculating again the power radiated by each dipole:

- $P_{D1} = 0.925 \times 0.2 = 0.185\ P_{in}$
- $P_{D2} = 0.925 \times 0.8 \times 0.859 \times 0.25 = 0.159\ P_{in}$
- $P_{D1} = 0.925 \times 0.8 \times 0.859 \times 0.75 \times 0.861 \times 0.33 = 0.135\ P_{in}$
- $P_{D1} = 0.925 \times 0.8 \times 0.859 \times 0.75 \times 0.861 \times 0.67 \times 0.863 \times 0.5 = 0.119\ P_{in}$
- $P_{D1} = 0.925 \times 0.8 \times 0.859 \times 0.75 \times 0.861 \times 0.67 \times 0.863 \times 0.5 \times 0.871 = 0.103\ P_{in}$
- $P_{TOTAL} = 0.701\ P_{in}$

The total lost power in the feed line is 0.299 P_{in}, which represents a 1.5dB loss. This is an improvement of just 0.3dB over the uniform strip feed line. This modest improvement probably doesn't justify the use of a more complicated to construct "optimal" feed line.

Matching network and balun

The dipoles and the feed line are structures intrinsically balanced but the input connector is unbalanced. Moreover, the impedance between the feed points A-B is 30Ω whereas the impedance of the SMA connector is 50Ω. Therefore, for best performance, an impedance transformer and a balun should be used between the feed points A-B and the input connector. The impedance transformer employed was a simple quarter-wave line section with appropriate characteristic impedance. The balun was based on the coupling effect between a copper strip - with one end soldered to the ground (reflector) and the other left open – and the grounded conductor of the quarter-wave transformer (Fig 63).

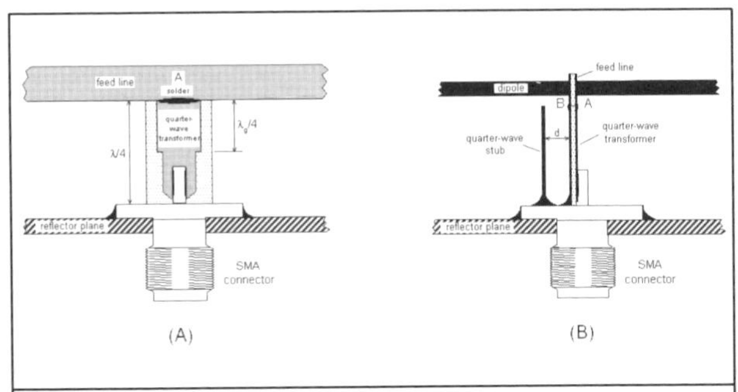

Fig 63: Quarter wave transformer and balun.

The quarter-wave impedance transformer is made from an FR-4 substrate symmetrical microstrip {see footnote 2}. The quarter-wave section is 5.25mm wide and 6.25mm long, with characteristic impedance equal to $Z_0 = \sqrt{50 \times 30} = 38.7\Omega$. The upper end of the quarter-wave section is soldered to the feed line. The lower end is connected through a 3.75mm wide 50Ω section to the SMA connector. The conductor on one side of the 50Ω section is soldered to the pin of the SMA connector and other side is soldered to the ground. Note that the feed point B is connected to the ground through the matching network. However, the path is λ/4 long, so the impedance from B to ground would be infinite if there were no radiation. The quarter-wave stub placed beside the matching network works in a way similar to the "bazooka" balun. The current induced in the stub cancels the effect of the current that flows on the external face of the

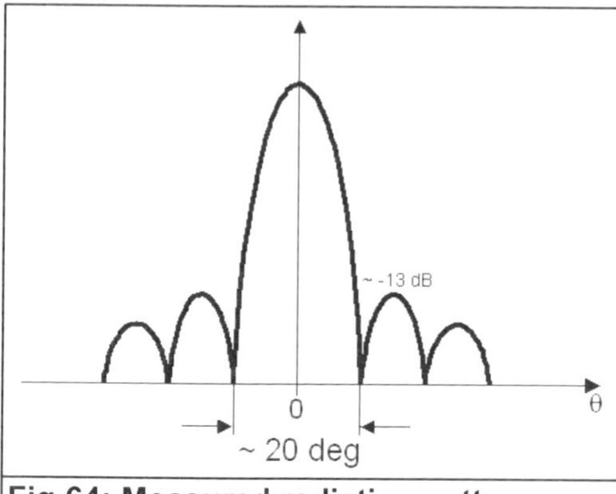

Fig 64: Measured radiation pattern.

microstrip conductor from B to ground. The distance d between the stub and the microstrip can be used to tune the input impedance for best matching.

Gain estimation

The gain of the antenna array can be estimated by several methods. The EZNEC antenna simulator gives 7.65dBi as the gain of an individual dipole over a conductor plane. As there are 10 such dipoles, the gain of array is 10 times greater, if the coupling between dipoles and the feed line losses are ignored. That would give an array gain equal to 17.65dBi.

Another way to estimate the gain is through the –3dB beam width. A formula that considers that side lobes follow a Chebyshev distribution 15 - 20dB below the main lobe amplitude, gives the gain as:

$$G = \frac{32600}{\theta_1 \times \theta_2}$$ where θ_1 and θ_2 are in degrees

According to the EZNEC software, the directivity in E plane is 67° (given by the dipole radiation pattern), whereas the directivity in H plane is 10.1° for $L/\lambda = 5$ (see Table 1). The gain calculated from the formula above is 16.8dB, that is, about 0.8dB lower than the first estimation.

Measurements

The measurement of the radiation pattern requires an anechoic chamber or a reflection-free environment. Sadly, none facility was available, so the pattern measured wasn't very precise. Nonetheless, the first nulls and the amplitude of the first side lobes were reasonably close to what was expected (Fig 64).

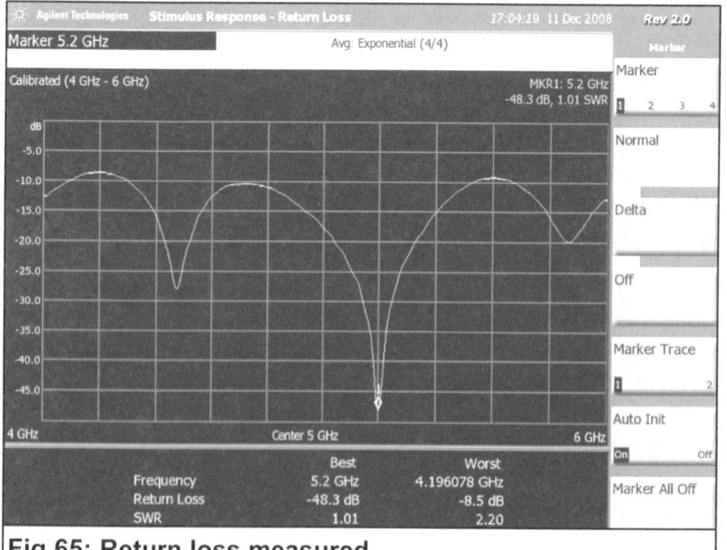

Fig 65: Return loss measured.

Fig 65 shows the return loss measured at the SMA connector. The bandwidth for RL equal to 20dB (1.22 SWR) is nearly 280MHz, but RL is better than 10dB (1.9 SWR) for almost all frequencies between 4 and 6GHz.

Bidimensional array

If larger gain or more directivity

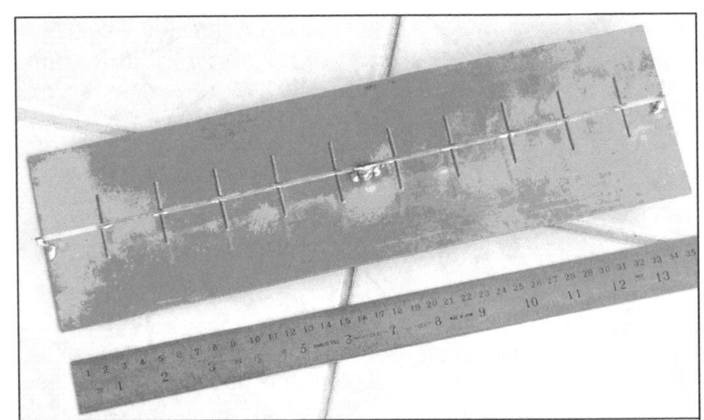
Fig 66: Picture of the prototype antenna array.

in the yz plane is required, the linear array can be expanded into a two-dimension array as shown in Fig 67. For instance, if the linear array is replicated 8 times in the y direction, the gain will increase by 9dB and the estimated theoretical gain for the array will grow to 26dBi. A point important to remember is that the bidimensional antenna array will keep the same low profile of the linear array. The linear sub-arrays must be fed in phase by an appropriate feed network, which can be constructed, for example, by using the same idea of the strip line feed for the individual dipoles. Another idea is using coax to feed the sub-arrays. An even larger gain can be achieved if the array is also replicated in the x direction. For example, duplication in the x direction increases the gain by 3dB and the expected theoretical gain for the entire array would be about 29dB.

Of course, as the antenna array is expanded the gain increases at the expense of the simplicity. There is a point when the parabolic reflector becomes more interesting, especially if the goal is only very high gain, without much concern about antenna profile.

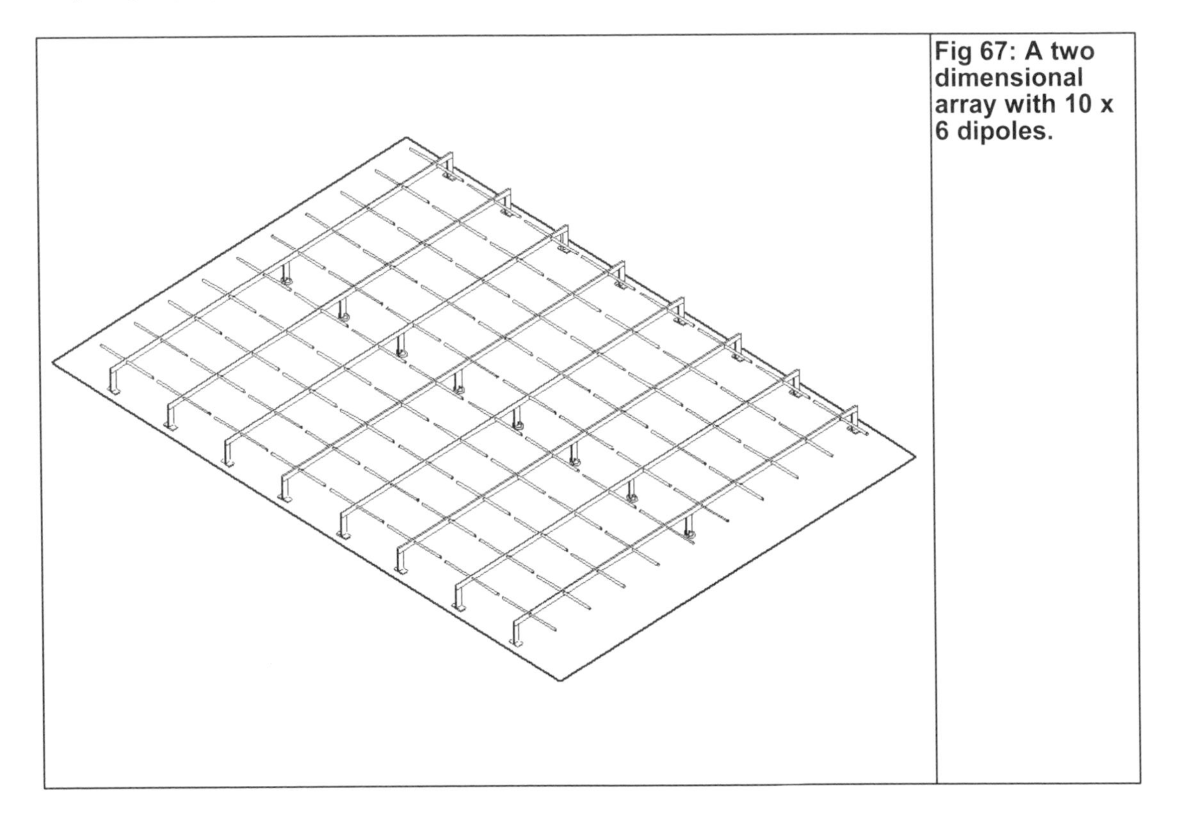
Fig 67: A two dimensional array with 10 x 6 dipoles.

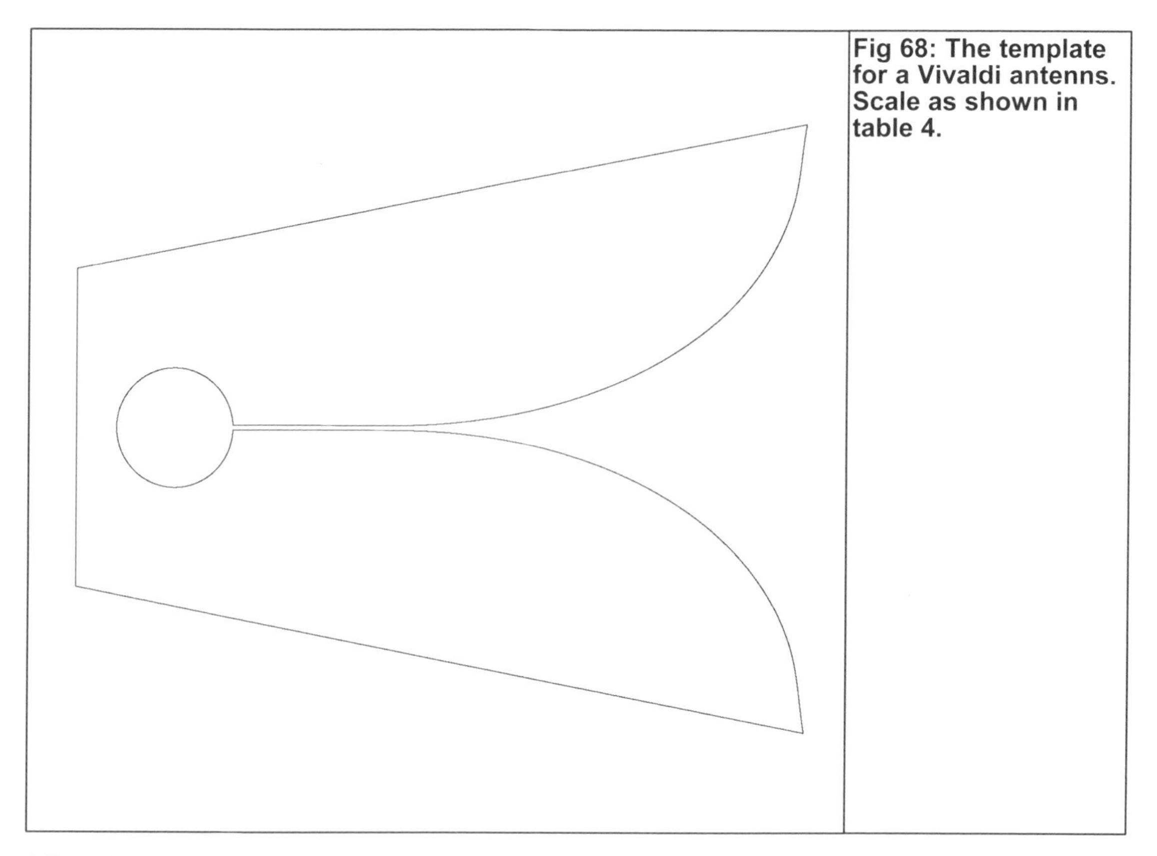

Fig 68: The template for a Vivaldi antenns. Scale as shown in table 4.

Vivaldi Antennas [11]

Vivaldi antennas are an exponential antenna from the same family as V-Beams and Rhombic antennas. They have exceptionally wide bandwidth. The lowest frequency is determined by the width of the opening.

The highest frequency is determined by how accurately the slot is formed. As an example the 75mm PCB version has an excellent Return loss from 5GHz to 18GHz and is usable from 2GHz to 22GHz.

All versions start with the template shown in Fig 68. Just take Fig 68 over to a copier and enlarge or reduce to the desired frequency range as shown in Table 4. Cut out the template and mark your material as shown in Fig 69. Thin Brass, Tin Plate, or PC Board has all been used. Now cut out your antenna. The Coax shield is soldered to one side of the slot and the coax centre conductor to the other side of the slot as close to the circle as practical as shown in Fig 70. Both semi-rigid and Teflon braided coax types have been used.

Table 4: Scaling for the Vivaldi antenna template.

Opening	Low End Frequency Response
40mm	10GHz
75mm	5GHz
150mm	2GHz
200mm	1GHz

Uses

Vivaldi antennas make excellent test antennas for use with a signal generator,

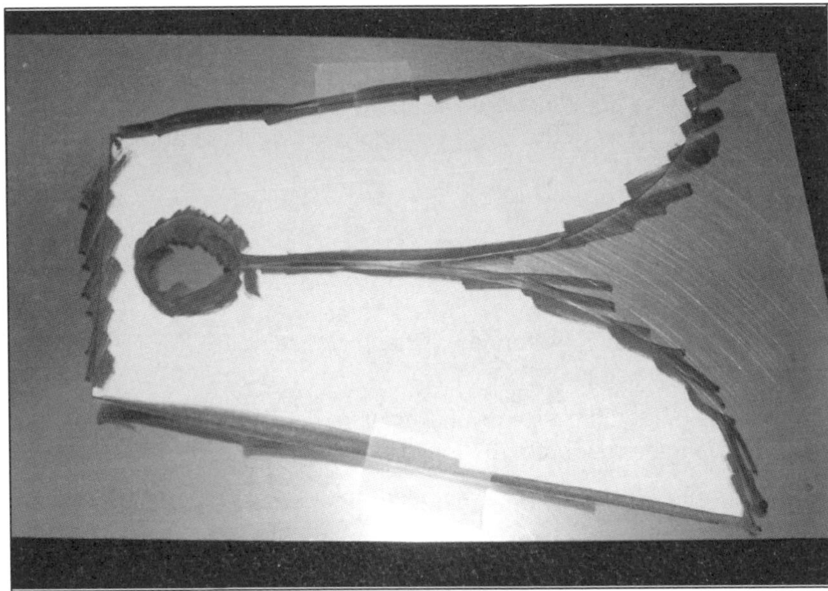

Fig 69: Marking out a Vivaldi antenna using the template placed on your chosen material.

spectrum analyser, or frequency counter. They also make a reasonable dish feed allowing the dish to work over several bands. The phase centre of the Vivaldi does move back and forth in the narrow region of the slot, but when the dish is focused at the highest frequency of planned use, the lower bands will be very close. Thus mount the narrow area of the slot at the focus of the dish.

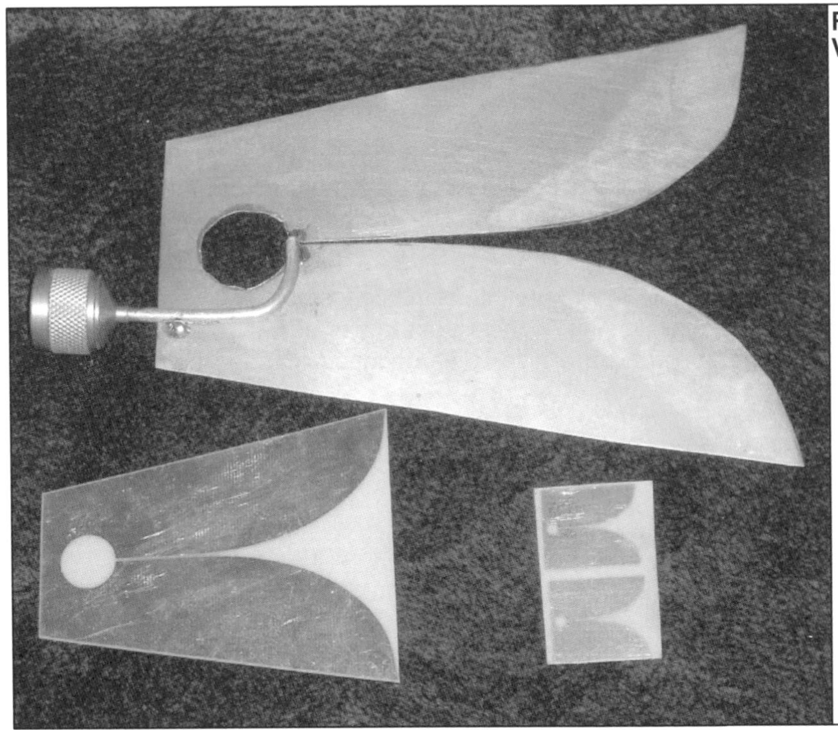

Fig 70: A completed Vivaldi antenna.

Next Step

If you wish to experiment with your Vivaldi, there are several ways to optimise the antenna over a narrow range of frequency. Moving the coax forward will improve response in the low end of its band. Using metal tape to block off parts of the circle will improve response at the high end of its band.

References

[1] Weatherproof UHF & microwave cavity antennas, Matjaz Vidmar, S53MV, VHF Communications Magazine 4/2009 pp 203 - 231.

[2] A. Kumar, H. D. Hristov: "Microwave Cavity Antennas", Artech House, 1989, ISBN 0-89006-334-6.

[3] H. W. Ehrenspeck: "A New Class of Medium Size, High Efficiency Reflector Antennas", IEEE Transactions on Antennas and Propagation, March 1974, pp. 329 - 332.

[4] Matjaz Vidmar: "An Archery target Antenna", Microwave Journal, May 2005, pp. 222 - 230.

[5] Antenna Array for the 6cm band, José Geraldo Chiquito, VHF Communications Magazine 1/2009, pp 37 - 53

[6] Kraus, John D, Antennas, second edition, chapter 3

[7] Skolnik, Merril: Radar Handbook, second edition, chapter 7

[8] Barter, Andy, G8ATD: International Microwave Handbook, 2002

[9] The ARRL UHF/Microwave Experimenter's Manual

[10] The ARRL Antenna Book, 20th edition

[11] Kent Britain 2E0VAA/WA5VJB, Kent Electronics – www.wa5vjb.com

Footnotes

{1} The double word strip line is used in this article to designate and differentiate the transmission line made of a strip cut from a double clad FR-4 board from the single word stripline, which has the usual meaning of a transmission line made of a flat strip of metal which is sandwiched between two parallel ground planes.

{2} Usually, a microstrip is a transmission line where one of the conductors is "live" and the other is a ground plane. In the quarter-wave impedance transformer used in the project, the conductors on both sides of the substrate are "live". The conductor printed patterns are the same on both sides.

Chapter 2

Power Amplifiers

In this chapter :

- 1W power amplifier for 9 - 11GHz
- A 45W amplifier for 23cm

odern power transistors have changed the way that we can generate useful output power on the microwave bands. This chapter contains two examples using modern devices.

1 Watt power amplifier for 9 to 11GHz [1]

Years ago many 10GHz amateur band amplifiers were described, all of these were narrow band. These amplifiers must be adjusted on the required 3cm transmit frequency. Wide band, integrated amplifiers have now been developed that can deliver high power. This article describes a wide band amplifier for the range from 9 to 11GHz with 30dB gain (±1dB) and a power output of 1W. The amplifier is suitable for both the 3cm amateur band and as a laboratory measuring amplifier because of its high gain and wide bandwidth.

Wide band amplifiers are substantially more complicated and difficult to design than narrow band amplifiers that are only operated a specified frequency. However they are more universal and can be used for most diverse applications.

Matching circuits for broadband amplifiers are often tapered striplines or a gradual impedance transformation made from several quarter wave transformers around the frequency response to reduce different forms of negative feedback, like resistive shunt feedback, e.g. reduce the gain. The transformation formulae can be computed with Tschebyscheff polynomials.

The use of an integrated amplifier that is already matched to 50Ω at the input and output is substantially simpler. The manufacturer Hittite [2] has such an amplifier, the HMC487, for the frequency range from 9 to 12GHz. It is only $5mm^2$ so cooling requires special attention, it must be soldered onto a printed circuit board and a large housings used. The gain is about 20dB and power output over is 1W. If the amplifier is made a two stage amplifier using an HMC441 as a driver, it makes a broadband amplifier from at least 9 to 11GHz with 30dB gain.

Initial consideration

The performance curves for 25°C in the data sheet should not be used to estimate the achievable gain and power output. This is because the amplifiers convert a considerable amount of power into heat, which is the price paid for the wide band performance. The HMC441 driver gives about 14dB gain and a power output of approximately 17dBm (50mW) at 10GHz, sufficient to drive the HMC487. The latter gives about 16dB gain and 31dBm (something over 1W) power output at 10GHz at a somewhat higher temperature. The gain data is for the 1dB compression point.

For an application that requires high linearity (e.g. SSB or D-ATV) one should remain somewhat

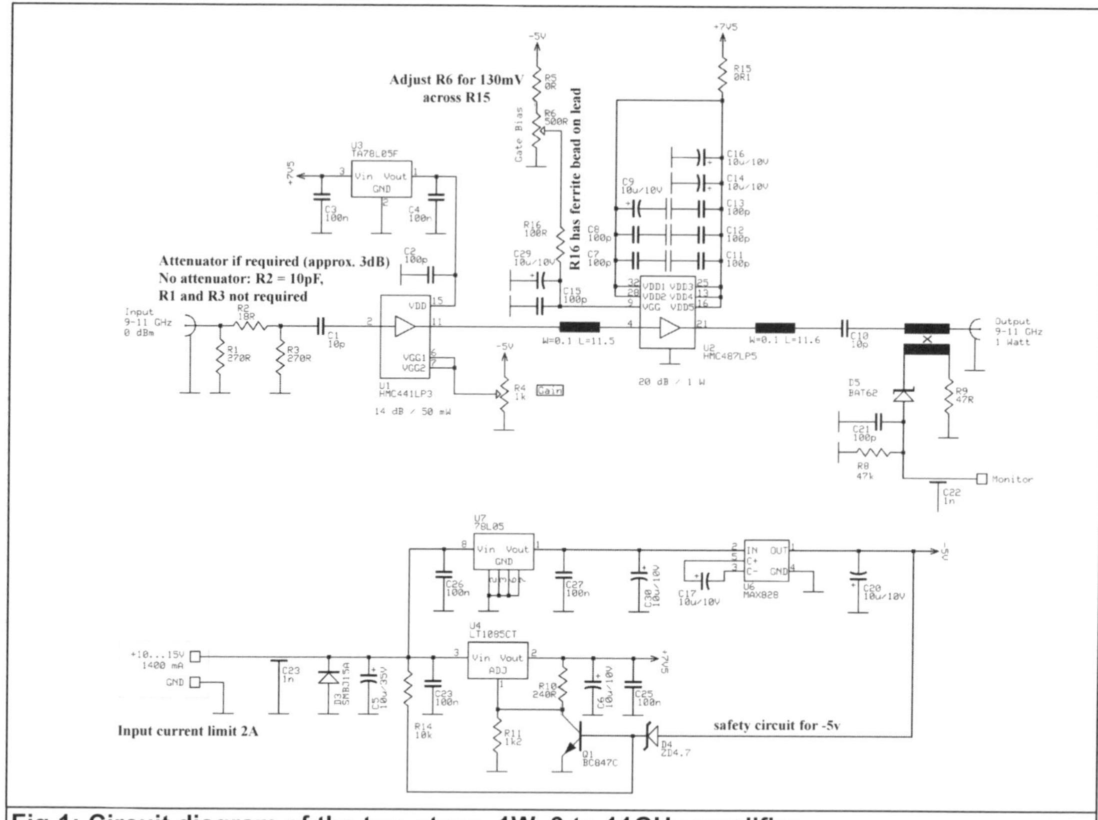

Fig 1: Circuit diagram of the two stage 1W, 9 to 11GHz amplifier.

under that for safety's sake, for FM (e.g. FM-ATV) one can go up to the saturation point, which should be between 1 and 2W. If the losses of the printed circuit board, the coupling capacitors and the transitions from the printed circuit board to the connectors is considered, the amplifier should supply an overall gain of approximately 30dB and 1W power output.

Circuit Description

The circuit diagram of the amplifier is shown Fig 1. There is an attenuator at the RF input made from R1 – R3, this is used when the application requires, if the attenuator is not required simply replace R2 with at 10pF coupling capacitor and omit R1 and R3. The first amplifier stage uses an HMC441 (U1), it can operate in the frequency range between 6.5 and 13.5GHz. In the frequency range selected here (9 to 11GHz) the return loss at the input is about 15dB; which is perfectly sufficient. The IC has its own 5V supply from U3. The trimmer R4 can be used to reduce the gain of U1 by up to approximately 5dB to be able to adjust the power output for use in certain systems. Under normal conditions this is not necessary, the trimmer was however fitted since sufficiently space was available.

The second amplifier stage follows using an HMC487 (U2). The IC is matched for 50Ω, a simulation on the PC showed that this can be improved with a small impedance transformation. At 10GHz S22 is improved from the original 5dB to approximately 10dB. Striplines at the input

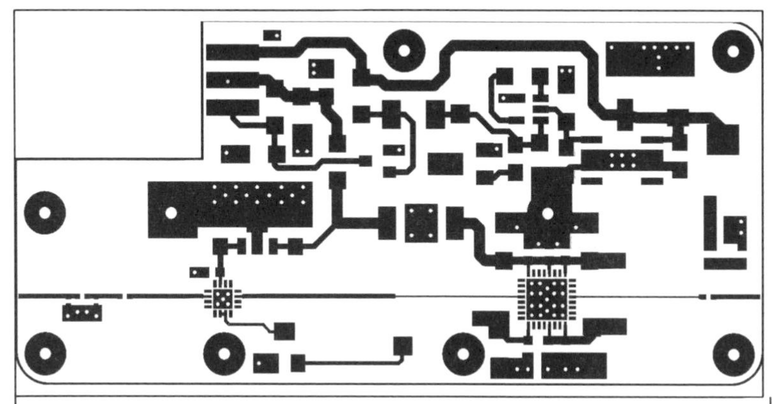

Fig 2: PCB layout for the 1W, 8 to 11GHz amplifier.

and output are used for this transformation.

The amplifiers already contain coupling capacitors. The coupling capacitors C1 and C10 were nevertheless planned since they have a substantially higher voltage rating and so they make the amplifier somewhat more durable. The integrated capacitors can easily become defective with inappropriate use, which leads to DC voltage at the input or output. The disadvantage of the coupling capacitor at the output is that it costs about 0.2dB attenuation. The PCB also has a directional coupler to give a monitor output.

The output stage has a small resistor of 0.1Ω (R15) to measure the supply current (100mV/A). The bias supply to the gate of U2 is fed via a 100Ω resistor (R16) with a ferrite bead on the lead, this prevents oscillation at the gate. The trimmer R6 adjusts the operating point of U2 by adjusting the negative voltage on the gate.

The external supply voltage (+10 to 15V) is fed via the feedthrough capacitor C23. An SMB J15A diode (D3) protects the amplifier against voltage peaks (transient ones). The supply for the output stage (U2) comes from the LT1085CT adjustable voltage regulator (U4). R10 and R11 set the supply voltage to 7.5V. The transistor Q1 protects the output stage if there is no negative gate bias, the supply for U4 is reduced to 1.25V. The voltage regulator U7 in conjunction with the charge pump U6 produces the negative gate bias. The MAX828 from Maxim [4] supplies the higher output currents needed for the gate of the output stage better than the familiar 7660.

Construction and alignment

The amplifier was developed on the proven microwave substrate RO-4003C from ROGERS [3]. As printed circuit board thickness 0.203mm was selected. This thin material makes very narrow 50Ω track widths possible and thus a better transition to the pads of the amplifier ICs. The thin material also has a thermal advantage because the high heat dissipation only has to be conducted through a thin board The PCB layout is shown in Fig 2 and the component layout in Fig 3.

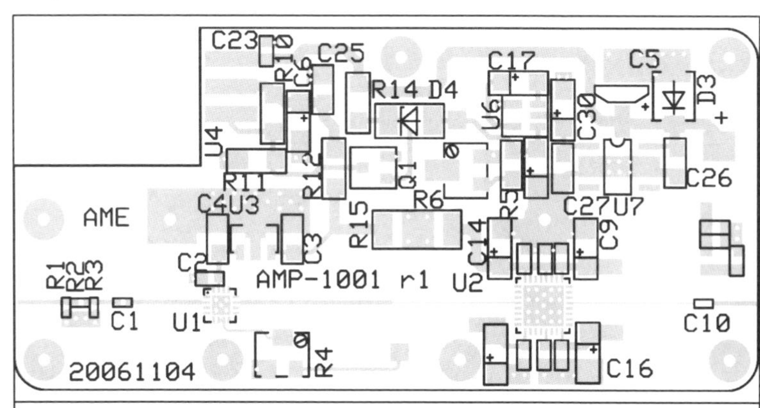

Fig 3: Component layout for the 1W, 9 to 11GHz amplifier.

The PCB must be plated through under the output IC in order to dissipate the approximately 9W of heat en-

Table 1: Parts list for the 1W, 8 to 11GHz amplifier.

D3	SMBJ15A	C1, C19	10pF, 0402, ATC [5]
D4	ZD 4.7	C2, C7-8, C11-13, C15, C21	100pF, 0603
D5	BAT-62 o.a.	C3, C4, C23, C25-C27	100nF, 0805
Q1	BC847C	C6, C9, C14, C16, C17, C20, C29, C30	10µF/10V tantalum SMD A
U1	HMC441LP3 [1]	C5	10µF/35V SMD Electrolytic
U2	HMC487LP5 [1]	C22, C23	1nF Feedthrough capacitor, M3
U3	TA78L05F	1	PCB from RO-4003C
U4	LT1085CT	1	Brass plate 2mm, milled
U7	78L05	1	Milled aluminium housing
U6	MAX828EUK [3]	2	SMA sockets with Microstrip tags
R1, R2, R3	attenuator if needed, SMD 0402		
R4	1kΩ trimmer, Bourns 3214W		
R5	0Ω 0805		
R6	500Ω trimmers, Bourns 3214W		
R16	100Ω fitted with ferrite bead		
R15	0Ω 2512		
R9	47Ω 0603		
R8	47kΩ 0603		
R14	10kΩ 1206		
R11	1.2kΩ 1206		
R10	240Ω 1206		
R12	0Ω 1206		

ergy. The IC (U2) is soldered to the PCB using flux and a hot air gun from the earth side of the PCB. The plated through holes should run freely with solder because this is the path to dissipate the heat, see Fig 4. Excess solder on the earth side of the PCB should be removed so that it does not prevent the PCB from laying flat.

The driver U1 is soldered in a similar manner with hot air. With this IC the plated through holes must be not completely filled, since only a small heat dissipation needs to be removed.

The entire PCB is soldered to a 2mm thick brass plate. For this, use an alloy with a lower melting point (Sn42Bi58) and a heating plate. The soldering procedure is carried out at 150°C. The brass plate gives faster heat distribution and thus better heat dissipation from the milled aluminium housing. Alternatively a milled brass housing can be used saving the transition to the brass plate just solder the PCB directly to the housing. However this is somewhat more expensive to manufacture, also the surface treatment of a brass housing e.g. gold, is substantially more expensive than a suitable plating for aluminium (e.g. chrome 6).

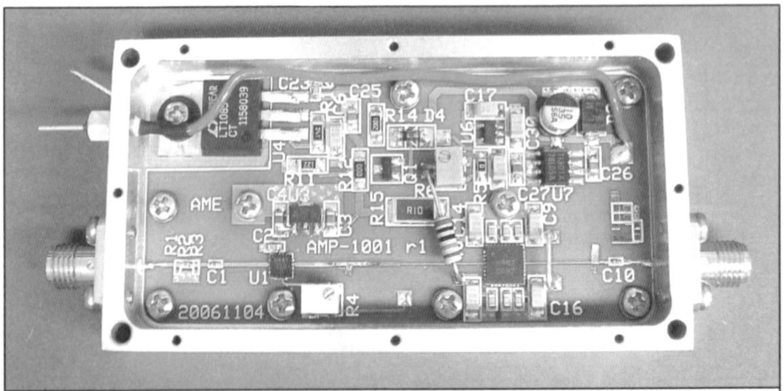

All further components, except R15 and R16, can now be fitted on the PCB. The supply voltage to the output stage U2 is made by a small cable link (see Fig. 4). Likewise a cable link must be used for the -5V

Fig 4: Picture of the completed 1W, 9 to 11GHz amplifier.

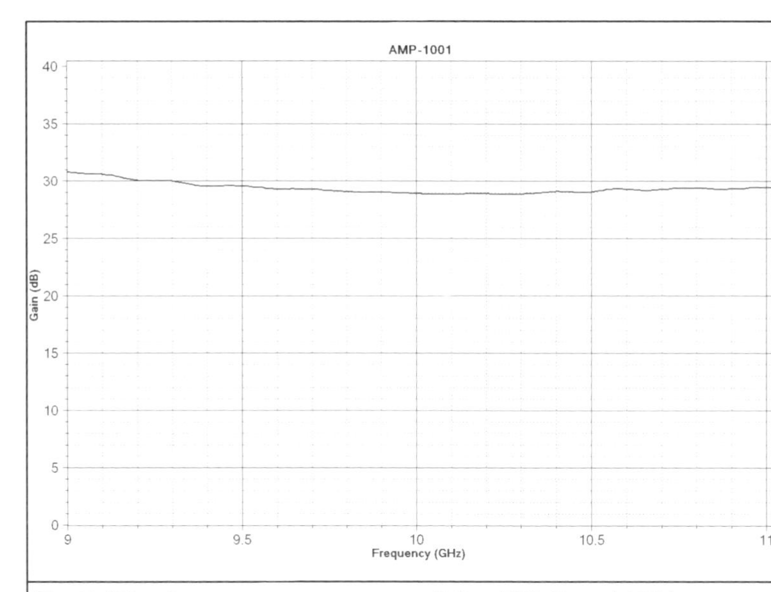

Fig 5: The frequency response of the 1W, 8 to 11GHz amplifier. Tha gain only fluctuates from 30dB by 1dB.

supply of R4. C10 must be a high quality microwave capacitor [6], otherwise the desired gain and power output will not be achieved.

For the alignment procedure use a supply with a current limit of 2A to protect the amplifier in the event of an error. Connect the supply voltage and check all voltage levels (7.5V, 5V, approximately -5V). Set a voltage of -1.8V at the connection to R16 using the trimmer R6. Switch off and fit R15 and R16. Switch the amplifier on and set the quiescent current of 1.3A (130mVs across R15) using R6. The amplification is checked using a network analyser. Small tags soldered onto the strip lines can be used to adjust the frequency response. Thus the amplifier is now ready for use.

The heatsink for the output stage should be generously rated (approximately 1W/°C).

Measured values

Fig. 5 shows the frequency response of the amplifier. The gain of 30dB was reached and varies only around ±1dB. The amplification can be increased by alignment with a tag to a maximum in the 10GHz region to approximately 34dB. Then there are larger fluctuations over the entire frequency range (9 to 11GHz). Power output (1dB compression) over the entire frequency range is somewhat more than 1W (typically 1.2W - 1.3W). The saturation output is about 1.5W.

The second harmonic at 20GHz could not be measured on the spectrum analyser. The efficiency as PAE (power added efficiency) is 9%. Thus the desired values for gain and power output required were achieved.

45W amplifier for 23cm [7]

LDMOS technology offers easy and economical microwave power. For many years I operated on 23cm with just 2 watts RF output but never experienced the aircraft scatter and consistent over-the-horizon contacts that I knew the band could support. Increasing the power by 13dB or so opened up the band in ways I didn't think possible. This article describes a 45W Class AB Power amplifier covering the 1.3GHz band that requires only a few watts of drive for full output. It is a modern, safe replacement for a single 2C39A valve amplifier and does not require a hazardous high voltage power supply. It is ideal for use with a DB6NT or similar transverter, a Kenwood TS2000X or similar transceiver, reaching a power level where the 1.3GHz band becomes interesting. The design is very simple and based around a single Freescale [8] MRF9045 28 volt LDMOS device. The PCB is designed to fit in to a readily available 72 x 55 x

Fig 6: A prototype 23cm power amplifier fitted to a small alluminium heatsink with fins running horizontally. A small black fan fives enogh airflow to keep the heatsink cool.

30mm tinplate box [9], and should be bolted to a suitable large, heatsink. Alternatively, a smaller heatsink with a fan can be used, as shown in Fig 6. A mini kit consisting of the PCB and MRF9045 is available [10].

This project uses small surface mount devices and requires some experience to construct and set up. It is therefore not suitable for novice constructors.

Circuit description.

Researching designs for the MRF9045 and similar devices it seemed that matching these devices to 50Ω at 1.3GHz requires a relatively simple microstrip network with low impedance (i.e. wide) lines on the input and output. Half a day playing with Appcad [11] and a Smith chart program [12] allowed me to get a reasonable handle on the required matching networks.

Fig 7 shows the complete circuit diagram. In order to reduce losses, the final circuit avoids using trimmers and shunt capacitors. Device input and output impedances are matched to 50Ω using series low impedance lines TRL2, 4 and 5. 'Shunt' lines facilitate the final tuning. The open circuit shunt lines TRL3, TRL6 perform the function of the shunt capacitors of lumped element designs and there is a short-circuited shunt line on the input, TRL1, grounded at the end by C5.

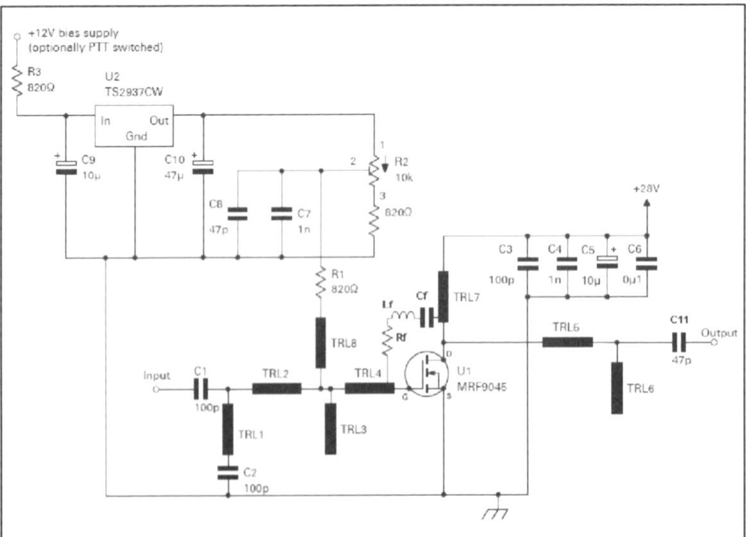

Fig 7: Circuit diagram of the 23cm amplifier.

Positive gate bias is fed via a similar high impedance line, TRL8, from a 5V regulator and potential divider to set the standing drain current to 350mA. The gate supply is decoupled by C7, C8 and C10. The connection to the gate is via a 'stopper resistor' R1. LDMOS devices are remarkably robust but they are susceptible to too much gate voltage. The only time I've destroyed a device during development is when I omitted this resistor. Without gate bias the amplifier takes very little current. Switching the 5V

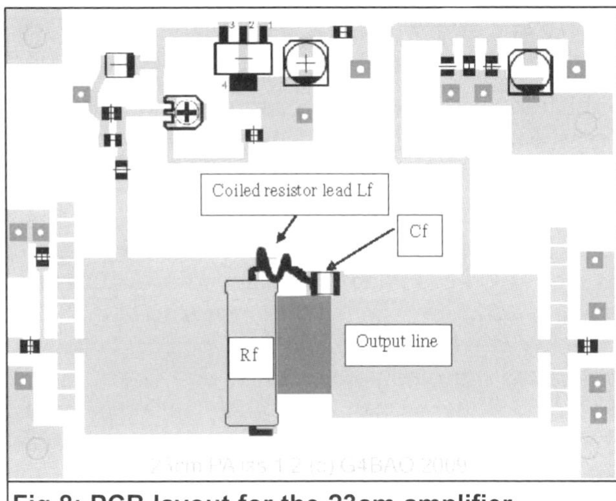

Fig 8: PCB layout for the 23cm amplifier.

regulator from the press to talk (PTT) line is a convenient way to switch the amplifier in and out of standby, minimising standby power consumption and heat generation. A small amount of inductively peaked negative feedback is applied between the drain and the gate of the FET. It consists of a 2W resistor and a very low ESR RF capacitor as a DC block. Winding two turns of the resistor leg around a 2.5mm drill makes up the inductance. This resistive feedback reduces the gain at lower frequencies while the added inductance reduces the feedback at 1296MHz. This is essential to the design; it vastly improves low frequency stability of the amplifier. Without this feedback there is a tendency for a spurious oscillation at around 220MHz when the amplifier feeds a short circuit or an open circuit with multiple quarter wavelengths of coaxial cable. It is apparent that, at 220MHz there must be a negative real part to the drain impedance, caused by Miller effect plus whatever impedance is presented to the gate.

Construction

The PA is built on a special 0.8mm thick low loss Teflon PCB, the overlay for which is shown in Fig 8. I developed the design using RF35 from Taconic Advanced Dielectrics Division in Ireland [13] to whom I gratefully acknowledge the sample of material that allowed the development.

Table 2: Parts list for the 23cm amplifier.

Component	Value	Type	Farnell reference
R1, R3, R4	820Ω	SMD 0805	9334890
R2	10kΩ	SMD preset	1173884
Rf	100Ω	2W carbon film	
C1, C2, C3	100pF	Np0 ceramic 0805	1414655
C8	47pF	Np0 ceramic 1206	1414734
C11	47pF	Np0 ceramic 0805	1414692
C4, C7	1nF	Np0 ceramic 0805	718567
C5, C9	100μF 35V	SMD electrolytic	8823120
C6	100nF	X7 ceramic 0805	9406387
C10	47μF 16V	SMD electrolytic	498762
Cf	Any value 240pF - 1nF		ATC100B
U1	MRF9045	LDMOS power FET	www.g4bao.com
U2	TS2937CW	SMD 5V regulator	7261365
TRL1-8	Microstrip matching lines on PCB		
Lf	Two turns of the Rf lead wound on a 2.5mm mandrel		

My first attempt at the design, using cheap FR4 material, worked after a fashion, but produced much less power and, more importantly, eventually failed, as the losses in the board caused the output microstrip line to heat up alarmingly and produce smoke after about 5 minutes.

Table 2 shows the component list for the amplifier. The only really critical component is the 1206 capacitor for the output coupling.

The MRF9045 used in the design is in a TO270 solder-down plastic package. These are now becoming more popular and eliminate the need for hazardous beryllium oxide in the package.

The device is soldered to a piece of copper plate approximately 27mm square and 1mm thick, which acts as a heat spreader between the MRF9045 and the heatsink. (Actually I used one face cut from a piece of waveguide 16). The spreader is soldered to the underside of the PCB and has two holes drilled in it to allow the spreader to be bolted to the heatsink.

Applications note AN1907 [14] from Freescale gives very strict guidelines on temperatures for soldering these devices but I thought, "what the heck, I can't afford a soldering oven, but I can afford to sacrifice a couple of devices if it doesn't work". So I tinned the underside of the PCB, the mating surface of the spreader and the topside where the drain and gate tabs go, bolted the spreader and PCB tightly together. I then dropped the module, spreader side down, on the kitchen ceramic hob, turned it on and waited for the solder to melt. It duly did, leaving a shallow 'pool' of solder in the rectangular hole in the PCB. I carefully placed the device into the hole, pressing down with the tweezers. I then quickly slid the module off the heat, keeping the pressure on the device until the solder set. Once the device is fitted you can solder in the other components. I have used this technique many times now with no ill effects on the devices. If you have a proper hotplate, all the better; I have now gone 'upmarket', using an old dry clothes iron, that I clamp upside down in a vice; that does the trick just as well.

Solder one end of the ATC capacitor, Cf, to the output line first, and "tin" the other end. The resistor, Rf, should be trimmed to match Fig 8. Ensure that the resistor stands clear from the PCB on its "leg". Solder the resistor short leg down first, then carefully bend the "coil" end such that it touches the unattached end of Cf. Make sure that the coil lead stays touching the capacitor, and does not "pull" on the end of the capacitor as it could break it once soldered. You can then solder it to the unattached end of Cf.

Make sure that the coil turns in the lead of Rf are not shorted, and that neither the capacitor, Cf, nor the coil shorts to ground, the screws or the PA transistor.

No special precautions or techniques are required for fitting the other components, apart from good microwave practice. Be particularly careful to avoid excess solder, whiskers or excessive flux residues as these can affect performance.

Alignment

Bolt the assembled board and heat spreader to a large heatsink. The input and output matching tabs are split so that you can lengthen or shorten them by bridging them with solder, if necessary. Start with all the tabs linked for maximum length:

- Connect the input from your 1.3GHz transmitter to the amplifier input via a suitable SWR meter.
- Connect the amplifier output to a power meter or dummy load capable of dissipating and measuring at least 50W.
- Connect the drain to 28 volts via an ammeter on the 1A range. Connect the gate bias supply, starting with minimum volts on the gate and very carefully increase the gate voltage until the device begins to take current. This onset is very sharp,

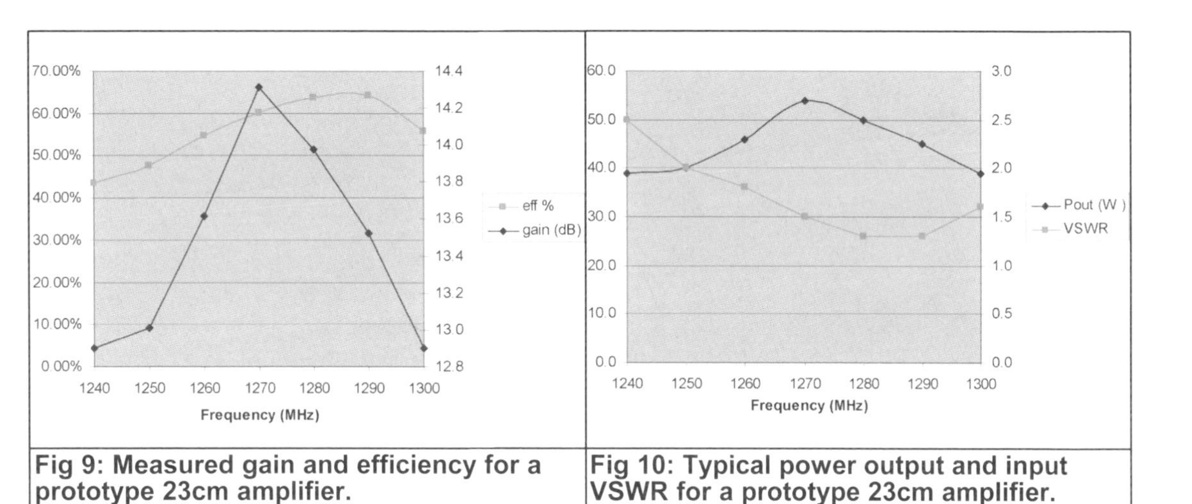

Fig 9: Measured gain and efficiency for a prototype 23cm amplifier.

Fig 10: Typical power output and input VSWR for a prototype 23cm amplifier.

so be very careful, as the drain current can easily swing up to many amps if you are not careful. Set the drain current to 350mA. Switch off power and then switch the ammeter to the 10A range. Switch back on and apply 0.5W drive and check the input VSWR. It should be better than about 1.7:1.

- If the input VSWR and output power are about right you're finished, so leave well alone. If not, switch off. Remove one section of one input tab at a time until the VSWR is less than 1.7:1 and check the output power and current. If necessary, trim the output tabs in the same way for maximum power. Do NOT try to make these adjustments with the amplifier operating. Turn up the drive in 3dB steps to 2W and check that the power increases about 3dB each time until it saturates. Now trim the output tab for minimum current consistent with maximum output power. It goes without saying, I hope, that you disconnect both the drain and source bias supplies before you trim the tabs.

Results

Test results for a prototype amplifier are shown in Fig 9 and 10.

Conclusions

This is an inexpensive PA and is relatively easy to build, with readily available components and produces a useful increase in output power for transverters and transceivers.

It is efficient and it is safe, in that it does not use any hazardous high voltage supplies. Note, though, it will give you a nasty RF burn if you are silly enough to poke your finger on the output line while transmitting.

It covers the whole of the 1.3GHz narrow band section and, with careful retuning (tabbing), could be adapted to cover the 1240MHz section of the band if band planning eventually forces a move of 1.3GHz narrowband lower in the band.

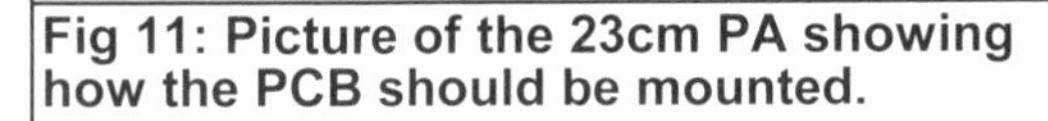

Fig 11: Picture of the 23cm PA showing how the PCB should be mounted.

Fig 12: Picture of the 23cm PA showing how the input connector should be fitted.

The amplifier can be made small enough to be masthead mounted, particularly if a fan is fitted, or with a bigger heatsink can be silent for shack use.

Note from the editor – G8ATD

Since the article was published in RadCom John has passed on some comments about construction that may be useful to others making this or similar power amplifiers. See John's web site [10] for any further updates.

When fitting the PCB in a tinplate box, solder the bottom ground plane of the PCB all round the base of the box, actually making it the base of the box, with the heat spreader level with the bottom as shown in Fig 11. The heat spreader can then be bolted directly to the heatsink with no further fittings required.

To ensure that it is level, before fitting the box, bolt the PCB alone to your heatsink, drop the box over the top and tack it to the PCB at suitable ground points on the top of the PCB. Unbolt and remove the PCB from the heatsink with the box now attached, check the heat spreader is level, then solder the underside ground plane all the way round.

Do not mount the SMA connectors directly on to the input and output capacitors as they can strain and crack them. Also there will be no room to fit a plug on them above the heatsink. Instead, mount them 6mm up the sides of the tinplate box and use very short links to the PCB. See Fig 12.

Do not fit the bottom lid of the tinplate box with the heatsink bolted on underneath because tinplate is a very poor thermal conductor and could cause the LDMOS device to fail due to overheating.

Use only M2.5 pan head screws or the screw heads will be dangerously close to the PCB tracks

Do not compromise on heatsink size, you have to get a lot of heat away from these devices

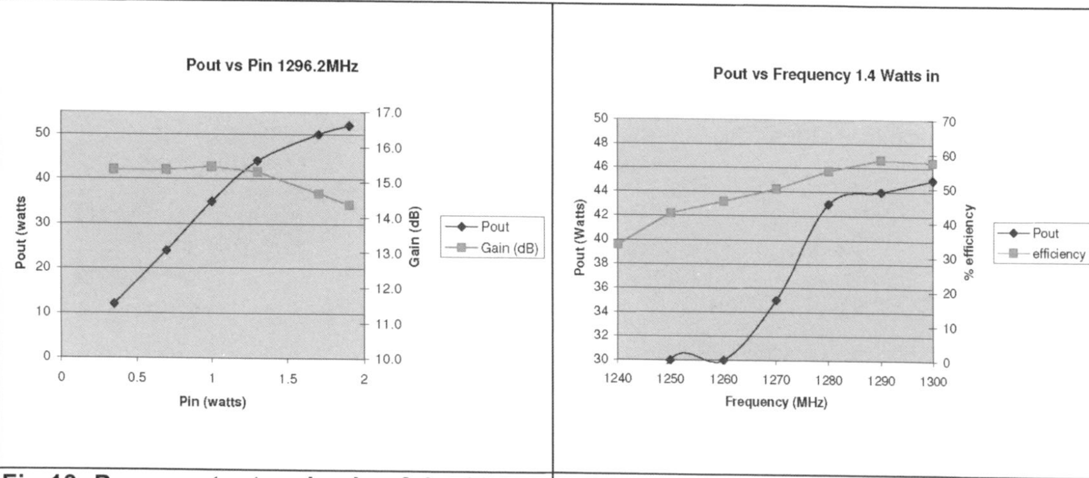

Fig 13: Power output and gain of the 23cm PA using a 5S9070 LDMOS device.

Fig 14: Power output and efficiency of the 23cm PA using a 5S9070 LDMOS device.

quickly, and a minimum size without a fan should be 120 x 100x 40mm.
Finally, John noted that the PA PCB has also been tested with MRF9060 and 5S9070 LDMOS devices, the results with the latter higher power device are shown in Figs 13 and 14.

References

[1] 1 Watt power amplifier for 9 to 11GHz, Alexander Meier, DG6RBP, VHF Communications Magazine 4/2007 pp 194 - 200

[2] Hittite Microwave corporation, 20 alpha Road, Chelmsford, MA01824

[3] Rogers corporation, 100 S. Roosevelt Ave, Chandler, AZ85226

[4] Maxim Integrated Products, Inc., 120 San Gabriel drive, Sunnyvale CA94086

[5] Hittite Application note "Thermal management For Surface Mounted Devices", Hittite Microwave corporation

[6] American Technical Ceramics, One the north Lane, Huntington station, NY 11746

[7] A 45W amplifier for 23cm, Dr John Worsnop, Ceng, MIET, G4BAO, RadCom June 2009 pp 63 - 65

[8] Freescale applications note AN1907, www.freescale.com/files/rf_if/doc/app_note/ AN1907.pdf

[9] Suitable tinplate boxes can be obtained from Alan Melia, G3NYK at www.alan.melia.btinternet.co.uk/componen.htm

[10] PCBs and MRF9045 devices will be available in Europe from www.g4bao.com, and in the Americas from www.wa5vjb.com.

[11] A free and fully functional copy of 'Appcad' can be downloaded from www.hp.woodshot.com/ appcad/appcad.htm.

[12] A demo version of 'Smith' can be downloaded from www.fritz.dellsperger.net/downloads.htm

[13] Taconic International, Advanced Dielectrics Division, Mullingar Business Park, Mullingar, Co. Westmeath Tel: 00353 (0) 44 9339071, e-mail: add@4taconic.com; www.taconic-add.com

[14] Freescale applications note AN1907, "Solder Reflow Attach Method for High Power RF Devices in Plastic Packages"

Chapter 3

Measuring Equipment

In this chapter :

- Improving harmonic frequency measurements with the HP8555A
- 10MHz - 10GHz diode noise source
- Noise factor measurement with older spectrum analysers

Radio amateurs cannot always afford the best measuring equipment but do need to be able to test their radio equipment to ensure that it complies with the modern emission requirements and that it is performing as well as it can to work that illusive DX. Some ideas in this chapter show how impressive measurements can be made on a radio amateur budget.

Improving harmonic frequency measurements with the HP8555A [1]

Sometimes there are small things that cause more trouble than they should

Approximately two years ago I wanted to examine the output spectrum of my VHF transceivers and used in ageing HP 141T spectrum analyser with the HP8552B IF module and HP8555A RF module. I was not absolutely content with the measurement results because I thought that my transceiver had poor harmonic suppression.

I searched for the cause of the allegedly poor harmonic suppression of my transceivers for many weeks. According to the display of the spectrum analyser the second harmonic was barely suppressed by 50dB.

Only after I had attached a 500kHz bandwidth cavity resonator filter to the output of the transceivers and the harmonic suppression was no better did I begin to doubt the result of measurement and to look for the cause in the spectrum analyser.

A study of the HP8555A module datasheet showed that a maximum of –40dBm at the mixer input results in a –65dB internally generated harmonic.

The self-noise of the HP8555A on the 1kHz range is -115dBm. For a reasonable execution speed the 300kHz range must be used on the 2GHz frequency range otherwise the reading is incorrect. The self-noise is then -92dBm. Thus the usable dynamic range of the HP8555A is -92dBm - -40dBm = 52dBm. The dynamic range can be improved by reducing the IF range, however theoretically this requires scan times of 100 seconds or more. Meaningful working (e.g. alignment of a lowpass filter) is therefore no longer possible. The HP8555A data shows this to be correct but I was not very happy with this result.

A detailed examination of the circuits showed that the HP8555A only uses a single diode mixer. Depending on the input signal amplitude the first mixer of the HP8555A can produce many mix products with the first and fourth harmonic of the local oscillator. There is no balanced diode mixer as used on other spectrum analysers.

The local oscillator is coupled to the mixer with a 10dB attenuator so the actual mixer diode is

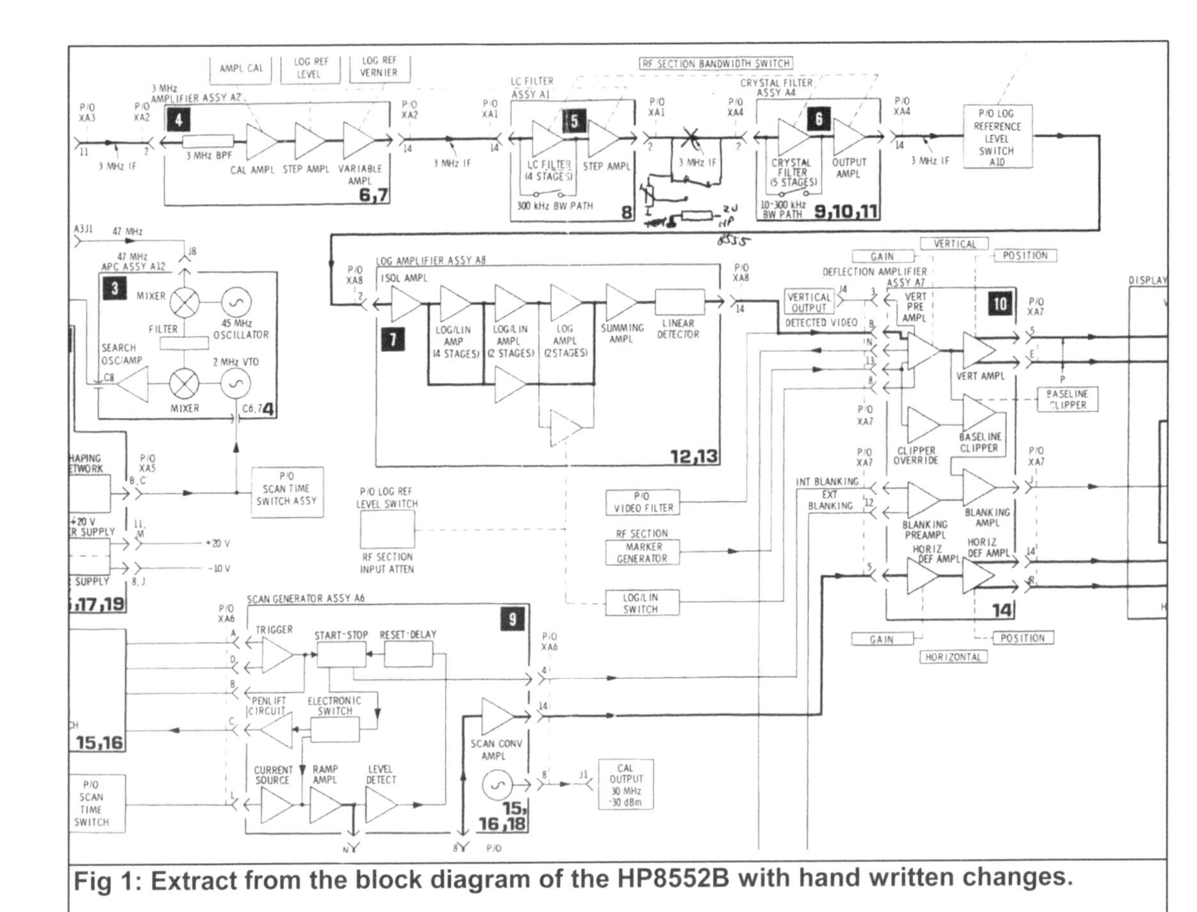

Fig 1: Extract from the block diagram of the HP8552B with hand written changes.

only fed with approximately +3dBm oscillator levels. With balanced diode mixers this is usually +13dBm.

The dynamic range of the HP8555A cannot be better because of the technique used. The successors HP8559 and HP8569, as well as the Tektronix 492 were developed using a similar concept and theoretically might be no better.

If the optional HP8444 tracking filters are used this problem does not arise because the tracking filter suppresses the fundamental by about 10dB if the harmonic appears in the IF filter. A tracking filter should always be used because ambiguities in the display are reliably avoided.

The solution

The only solution is using another input mixer, the following text describes this in detail.

A separate mixer with a downstream amplifier is used for the primary frequency range up to 1.8GHz. This is operated with a level of +13dBm at the local oscillator input. The following amplifier (Fig 3) raises the signal by approximately 13dB and is a wideband output for the mixer. Without this amplifier the new mixer would only work on the 2050GHz IF range with a bandwidth of approximately 50MHz due to the following 3 stage cavity resonator filter. It would be just as bad as the original mixer.

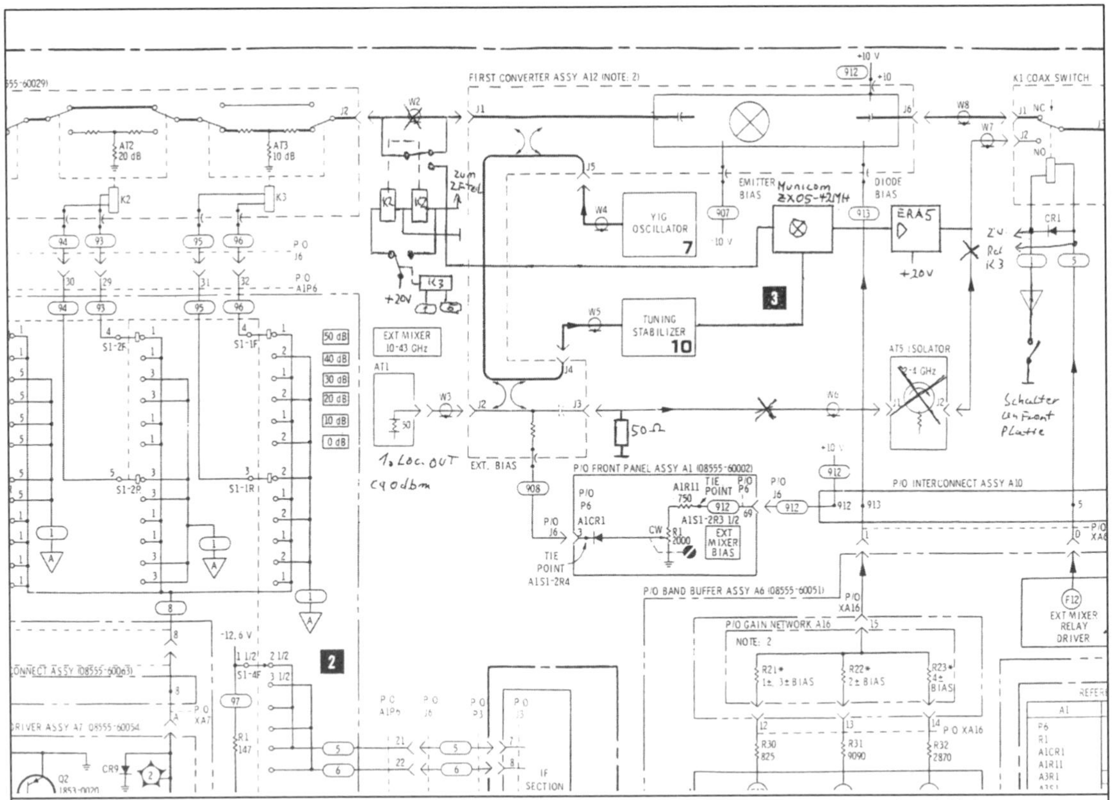

Fig 2: Extract from the block diagram of the HP8555A. The changes and additions are hand written.

Fig 3: The MMIC amplifier module that follows after the new mixer built into a tinplate box.

The new mixer is a three-way balanced ring mixer from Municom, type ZX05 42MH in a housing with 3 SMA sockets (Fig 4).

The amplifier is an ERA5 MMIC with approximately 18dB gain. It is operated with +20V and has a bias resistor of 270Ω. The redundant gain of approximately 10dB could be reduced at this point so that the overall gain is correct. It is better to attenuate the redundant 10dB in the IF amplifier before the crystal filter. This gives better control and 10dB more sensitivity.

This comes with the disadvantage of worse third order intermodulation in the first filter of the IF. Therefore two switchable outputs of mixer for the primary frequency range are used.

It is to be noted that with the additional mixer the second mixer is fed too highly therefore no modulation and intermodulation measurements are possible.

However the noise figure is better by approximately

Fig 4: The additional mixer is fitted under the metal angle. The connections: left to LO, middle IF output, right RF input

10dB. The original mixer should be used for narrow band measurements. Because of the small IF range, the dynamic range is better, because the noise decreases. The additional mixer should really only be used for harmonic frequency measurements because the intermodulation in the following mixer does not play a role. The change over switch is fitted on the front panel using the hole normally used for the external mixer bias potentiometer (Fig 5). The switch is used when high inputs would be applied to the original mixer. The input for the external mixer is not available following this modification; this socket now carries a 0dBm signal from the local oscillator.

The modifications

For the installation of the external mixer the following parts are required:

- 1 Municom ZX05-42MH mixer
- 1 Bistable coax relay for up to 18GHz
- 2 Miniature DPDT DIL 12Volt relay
- 1 Miniature SPDT switch
- 1 trimmer potentiometer 1kΩ.
- 1 SMA terminating resistor

The following aids are required:

- The original service documents for the HP8555A and HP8552
- An extension cable for operating the modules outside the chassis.
- Cable connection between RF and IF module.
- Special key for the SMA plugs this should be very slim and strong.

There are also the plug links for the modules to the basic equipment e.g. RS Electronics. There are Conec plug links to the other modules.

The AT5 isolator under the panel that carries the input mixer is redundant an can be removed. The 18GHz relay is fitted in its place. This connects the input attenuator with the two input mixers (Fig 6).

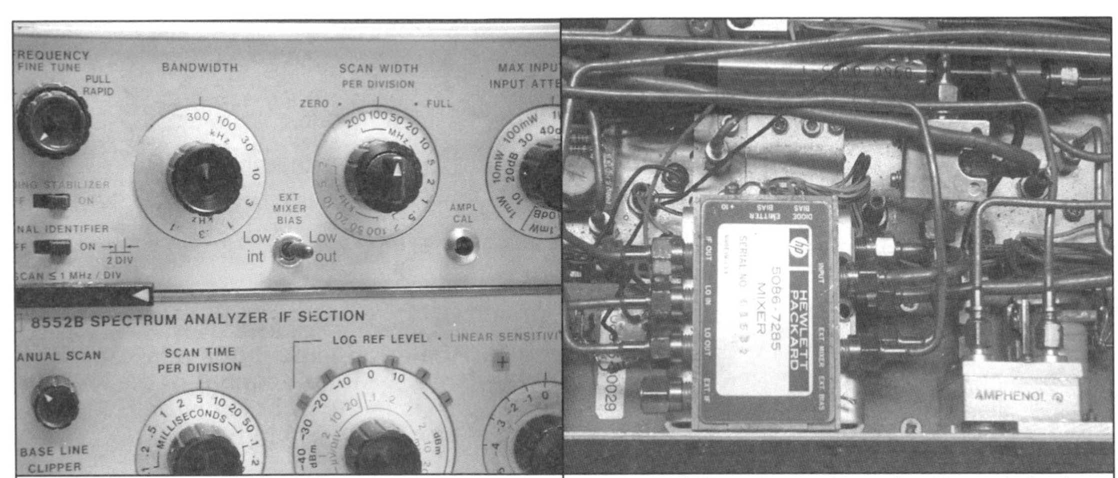

Fig 5: Arrangement of the new switch on the front panel.

Fig 6: 18GHz relays sit under the original mixer. The SMA terminating resistor can be seen on the left.

Relay K1 that was used to switch the intermediate frequency port to the external mixer input now selects the outputs of the two mixers.

The additional relay K3 has its coil in parallel with relay K1. +Pol on relay K1 is no longer connected to the -12.6V return but connected via the switch on the front panel to ground.

The change over switch in relay K3 supplies +20V to the 18GHz coax relay K2 and the relay in the IF port. The 18Ghz relay is a bistable version and has 2 coils that swap themselves over with internal switch contacts after the relay has been operated. This can be seen from the handwritten modifications in Fig 1 and 2.

A 50Ω SMA terminating resistor is fitted where the isolator connected to the original mixer. The original input for an external mixer now serves as output for a tracking generator,

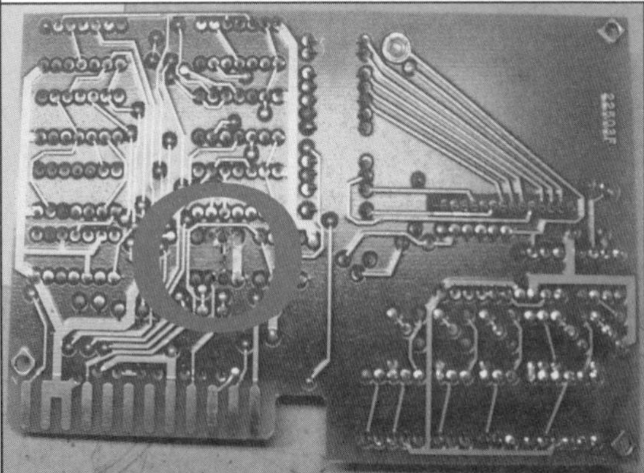

Fig 7: All RF connections are made using semi rigid coax with SMA plugs.

Fig 8: A connection is isolated on the back of the relay PCB.

Fig 9: A new connection is made on the front of the relay PCB.

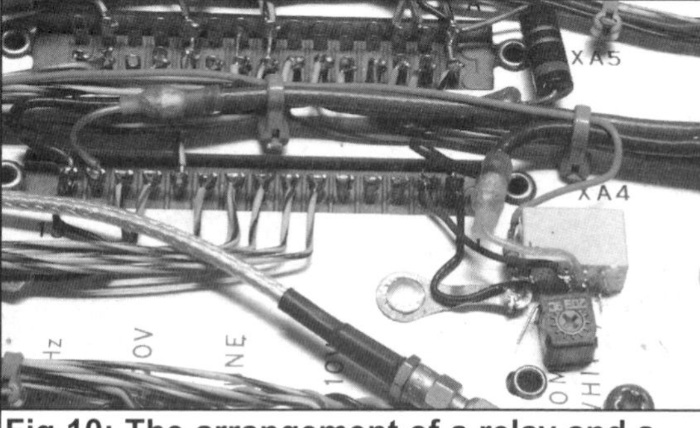

Fig 10: The arrangement of a relay and a potentiometer for adjustment of the level in the IF port is in the right lower half. Also see the block diagram in Fig 1.

however it has approximately 13dB lower level (now 0dBm). The original local oscillator output is stabilised by a module on the lower surface of the chassis, the signal from the local oscillator for the additional mixer is taken from here.

The existing semirigid cables can be used for connections between individual RF modules. If these are bent too much they no longer look beautiful and the screen can break open. In addition it can be difficult to fit the parts such as mixers and amplifiers in position (Fig 7). There are semirigid cables that can be use more easily. The relay PCB originally activates relay K1 if the spectrum analyser is switched to an external mixer. It is necessary for the relay driver to supply the relay only if the 10MHz to 1.8GHz range is selected. The tracks on the lower surface of the relay PCB are cut at IC U7 between pin 2 and pin 7 (Fig 8). A new connection is made on the topside of the PCB between pin 2, U7 and pin 8, U8; see Fig 9. In the original diagram the pin allocation of the IC U8 is drawn incorrectly.

The adjuster added in the IF (Fig 10) reduces the additional gain from the new mixer. Feeding a signal into the spectrum analyser and switching between the two mixers while making

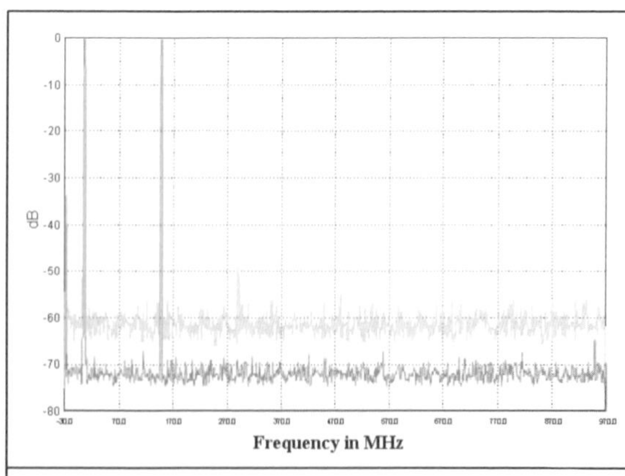

Fig 11: These curves show the difference between the original mixer and the new mixer. The conditions were identical for both curves.

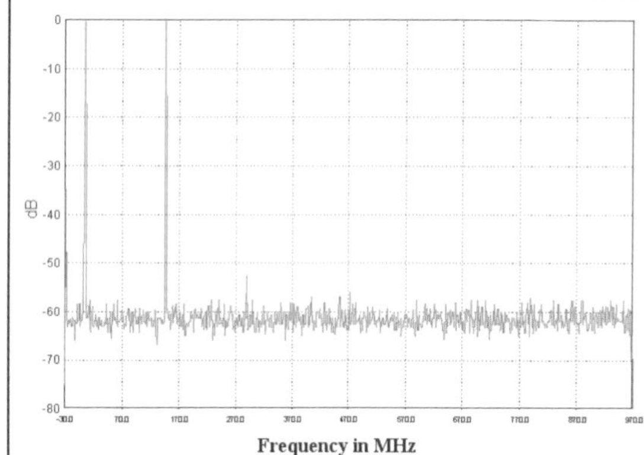

Fig 12: Detailed view of measurements with the original mixer.

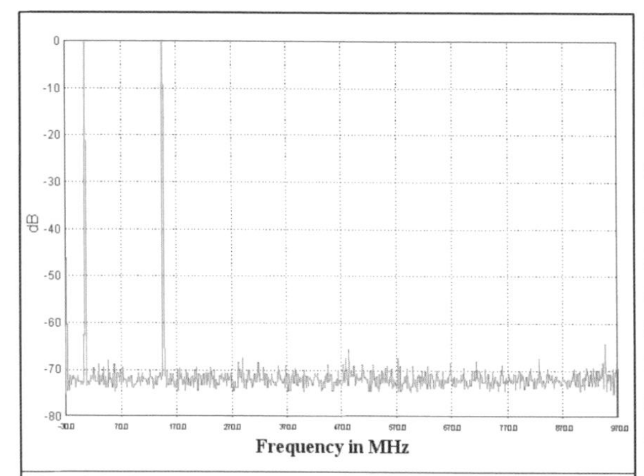

Fig 13: Detailed view of measurements with the new mixer.

adjustments until the two give the same display adjust this.

Thus the modification is finished.

Comparative measurement

The conclusion was a comparative measurement with an SMS2 signal source. The harmonics were suppressed with a lowpass filter and a cavity resonator filter connected between the signal source and the spectrum analyser. For the measurement accuracy the same settings were used on the spectrum analyser and the same input signal was used, only the mixer was switched. The measured curves are shown in Fig 11 to 13.

10MHz - 10GHz Noise source diodes [2]

In VHF Communications 1/2007 [3] I described a simple 10MHz to 3.5GHz diode noise source, the purpose of that article was to explain how to build a very simple noise generator using the NS-301 noise diode, either for applications like noise figure measurement or for a broadband noise generator for scalar applications with a spectrum analyser.

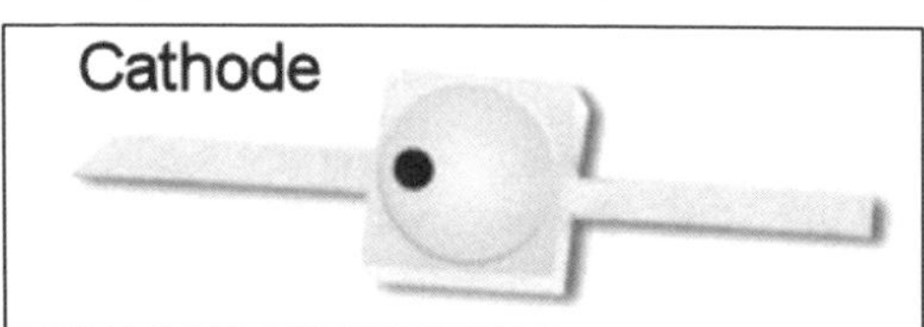

Fig 14: NS-303 noise diode.

Specification:

- **Case: Metal-ceramic gold plated**
- **Frequency range: 10Hz - 8GHz (max 10GHz)**
- **Output level: about 30dBENR**
- **Bias: 8 - 10mA (8 - 12V)**

Now I will describe a 10MHz - 10GHz noise source generator with an improved bias network that uses the NS-303 noise diode.

This project was born some months before the 13th E.M.E. (moon bounce) conference in Florence during August 2008, the organisation asked me to cooperate to build some noise source generators to give to participants during the conference.

Tests and measurements are supported by 20 pieces of noise source generators built for this conference, so I think that results are very reliable and repeatable.

Circuit diagram and components

The noise generator uses the NS-303 diode (Fig 14) that is guaranteed up to 8GHz but following the descriptions below it will be very easy to reach 10.5GHz making it possible to use it up to the 3cm band (10.368 GHz), using a diode of moderate price.

The aim of this article is to explain how to build a noise generator using easy to find components.

The circuit diagram, Fig 15, is very simple, the power supply is 28V pulsed AC applied to connector J1 which is normalised in all the noise figure meter instruments. U1 is a low dropout precision regulator to stabilise the voltage for the noise diode to 8 - 12V, the current through the

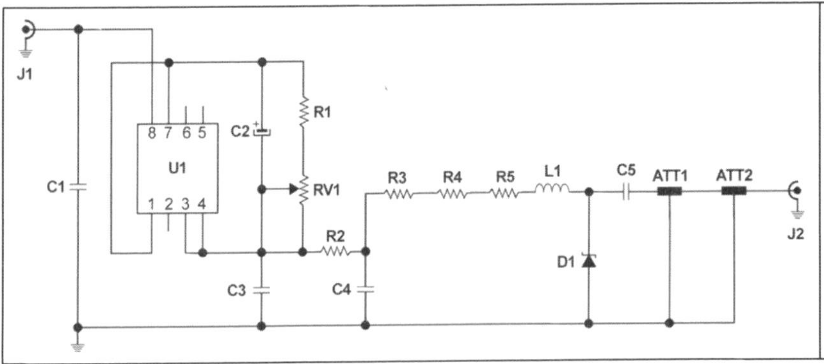

Fig 15: Circuit diagram of the 10MHz - 10GHz noise source using an NS-303 noise diode.

diode can be around 8 to 10mA set by trimmer RV1.

R3, R4 and R5 resistors

These resistors can be a total of 100 - 220Ω, the total value is not critical, the 0603 case is very important in order to keep the stray capacity as low as possible, it would be better to solder the resistors without using copper track on the PCB see Fig 17.

ATT1, ATT2 Attenuators

These attenuators are very important to obtain an output level of about 15dBENR but more important to obtain an output return loss as low as possible.

In my previous article in issue 1/2007 I described this concept very well, the mismatch uncertainty is the main cause of errors in noise figure measurement [4].

The total value of attenuators ATT1 + ATT2 can be around 14dB, the pictures in Fig 18 and 19 show a 6dB chip attenuator mounted on the PCB and a 7 or 8dB external good quality attenuator, in fact the output return loss depends mainly on the last attenuator (ATT2). The first attenuator (ATT1) can be less expensive and built directly on the PCB because it is less important for the output return loss.

I used a 7 or 8dB external attenuator in order to obtain the best output ENR value because every diode has it's own output noise.

Everyone can change the output attenuator depending on the ENR that is needed; in this project I chose an output level of 15dBENR so the attenuators have a value of 14dB.

Table 1: Parts list for the 10MHz - 10GHz diode noise source.

D1	NS-303 noise diode	J1	BNC fenmale connector
U1	LP2951CMX, SMD SO8 case	J2	SMA male panel mount SM24A Suhner 13SMA50-0-172
C1	10nf, 0805	R1	100Ω, 1206
C2	1µF, 25V tantalum	R2	18Ω, 0805
C3	100nF, 0805	R3, R4, R5	33Ω to 68Ω, 0603
C4	1nF, 0805 COG	L1	6.8 or 8.2nH 0603, BCG-6n8-A
C5	2 x 1nF, 0805 COG in parallel - see text	RV1	100Ω trimmer multi turn SMD POT-SM-101-M
ATT1	16dB chip attenuator DC - 12GHz	PCB	25N or RO4003 or RO4350 30mil, ε_r 3.40, 11 x 51mm - see text
ATT2	27 or 8dB external attenuator DC - 12GHz or better DC - 18GHz		

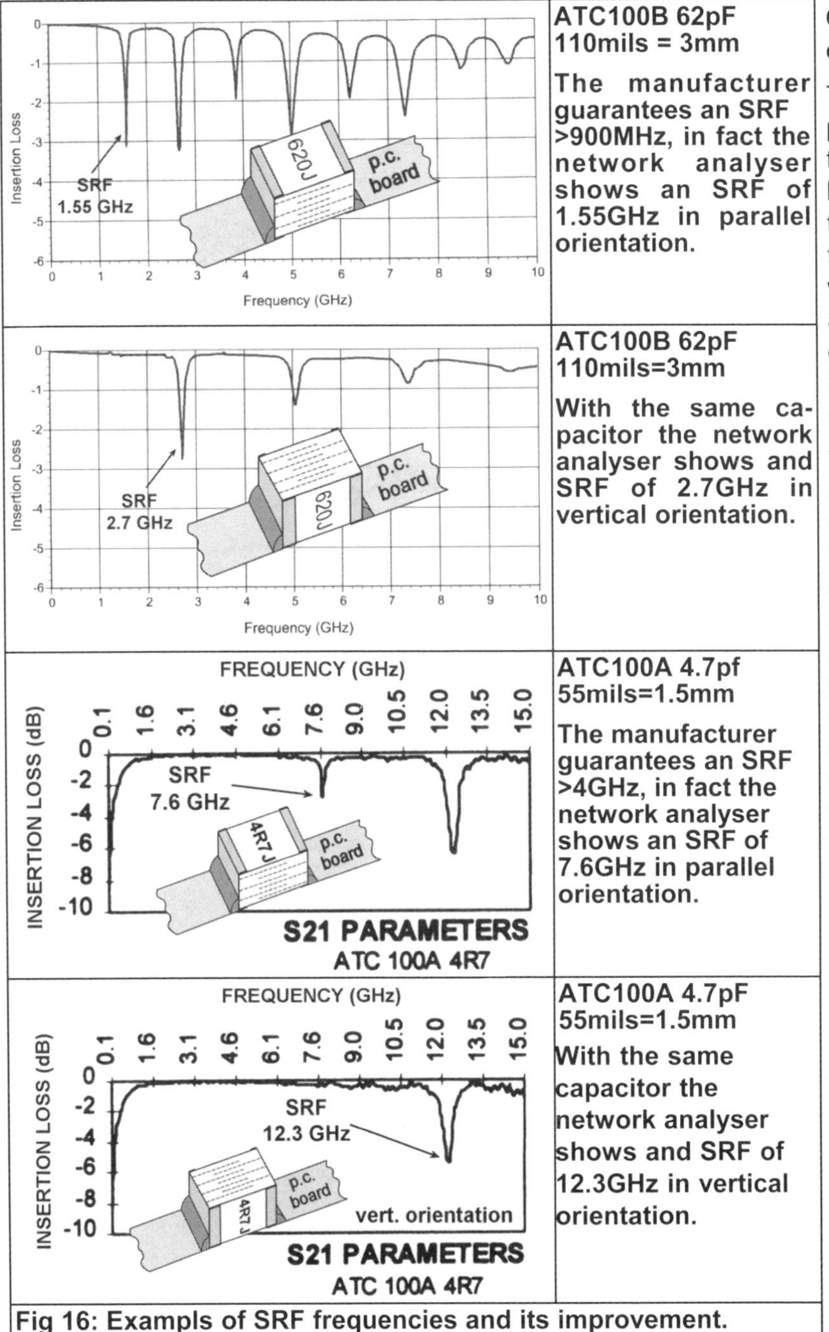

Fig 16: Exampls of SRF frequencies and its improvement.

C5 dc block output capacitor

The selection of this capacitor is very important to flatten the output level; in the previous article I only quickly mentioned this fact because we were only talking about 3.5GHz, now in order to reach 10.5GHz I will do a better description.

The dc block capacitors are used to block the dc voltage and to pass the RF signal with the minimum possible attenuation. If you use the ATC100A or 100B capacitors they have a very low insertion loss but have the problem of self resonance in ultra wide band applications, the graphs in Fig 16 show 2 examples how you can improve the SRF with vertical orientation.

Fig 16 shows 4 graphs of the SRF frequencies for ceramic capacitors and how to improve the SRF of ATC100A or 100B capacitors for ultra wide band applications.

My decision was to avoid ATC capacitors and to find some capacitors without any SRF and lower Q, after many attempts and researches I found that NP0 class 1 multi-layer capacitors with an 0805 case that have the best performance referred to low level applications (not to be used in RF power applications or in low noise amplifiers).

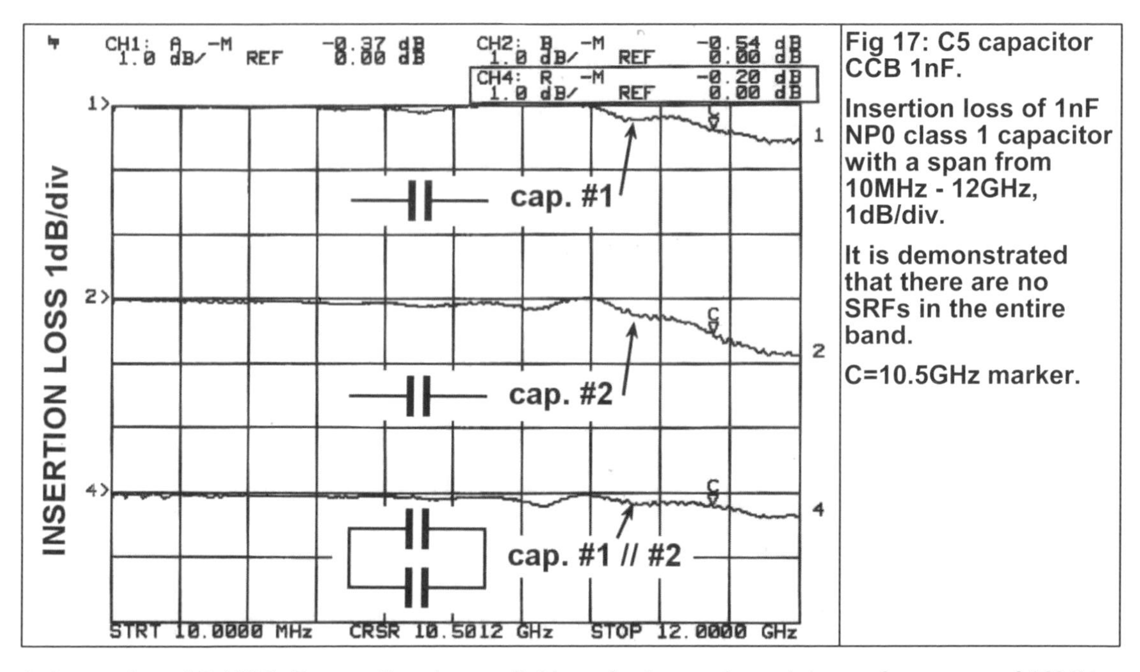

Fig 17: C5 capacitor CCB 1nF.

Insertion loss of 1nF NP0 class 1 capacitor with a span from 10MHz - 12GHz, 1dB/div.

It is demonstrated that there are no SRFs in the entire band.

C=10.5GHz marker.

I choose to put 2 1000pF capacitors in parallel in order to reach a minimum frequency of 10MHz.

For ultra broadband applications the ATC manufacturer has a capacitors of 100nF with 16KHz to 40GHz frequency operation in a 0402 case [5] but I prefer to avoid this special component and use more easy to find one.

In Fig 17 the 1nF capacitors show a low insertion loss, with 2 capacitors in parallel, the marker C shows an insertion loss of about 0.2dB at 10.5GHz that is appropriate for this project at low price.

PCB

The noise generator is considered a passive circuit so it is not necessary to use very expensive Teflon laminates, moreover the track length is so short that the attenuation introduced makes it unnecessary to use Teflon laminates. I selected ceramic laminate, that is very popular in RF

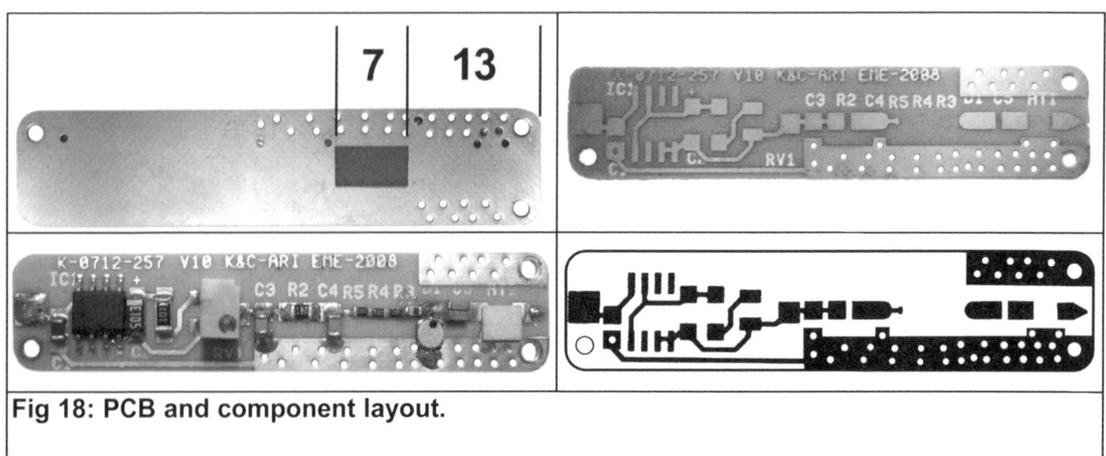

Fig 18: PCB and component layout.

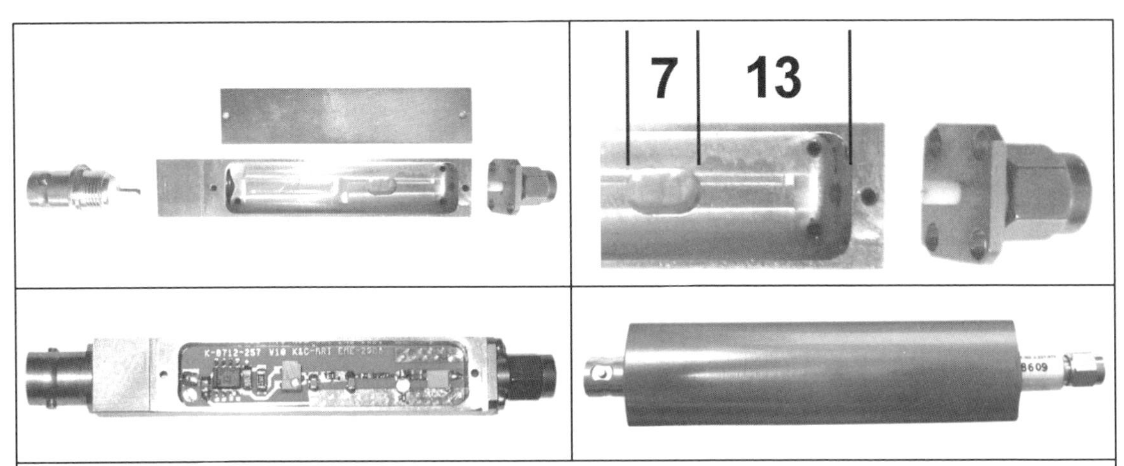

Fig 19: Box and final release.

applications, with ε_r = 3.40. It is available in several brands and they all have the same performance, Rogers RO4003 or RO4350, Arlon 25N etc..., with a thickness of 30mils (0.76mm).

In order to easily reach the 10GHz band it is necessary to remove the ground plane around R3, R4, R5 and L1, the size is 7 x 4mm (Fig 18)

Metallic box

As shown in Fig 19 the components of the noise source generator are enclosed in a very small milled box. Every box behaves like a cavity excited by several secondary propagation modes. For higher frequencies or in medium size boxes the RF circuit will also have many secondary propagation modes at various frequencies. Since every box is different in size, shape and operating frequency the calculation of secondary propagation modes is very difficult. To avoid this problem microwave absorbers are very often used placed into the cover of the box to dampen the resonance [6].

I selected a very small box in order to avoid both the secondary propagation modes and the microwave absorber; the size that I used gives no trouble up to 10GHz.

If someone wants to increase the size of the box (internal size) it will be necessary to use a microwave absorber.

It is also necessary to remove part of the ground plane in the metallic box by milling a 7 x 4 x 3mm deep slot corresponding to R3, R4, R5 and L1.

Bias current

The nominal current should be 8mA, during my tests I found that the output noise level has a quite strange but interesting variation: increasing the diode current the output noise level decreases by about 0.5dB/mA up to about 9GHz, beyond this frequency the effect is exactly the opposite.

Fig 20 shows the difference in output ENR of about 1dB with 8 and 10mA bias current and Fig 21 shows a little improvement of frequency range by about 500MHz with 8 and 10mA bias current.

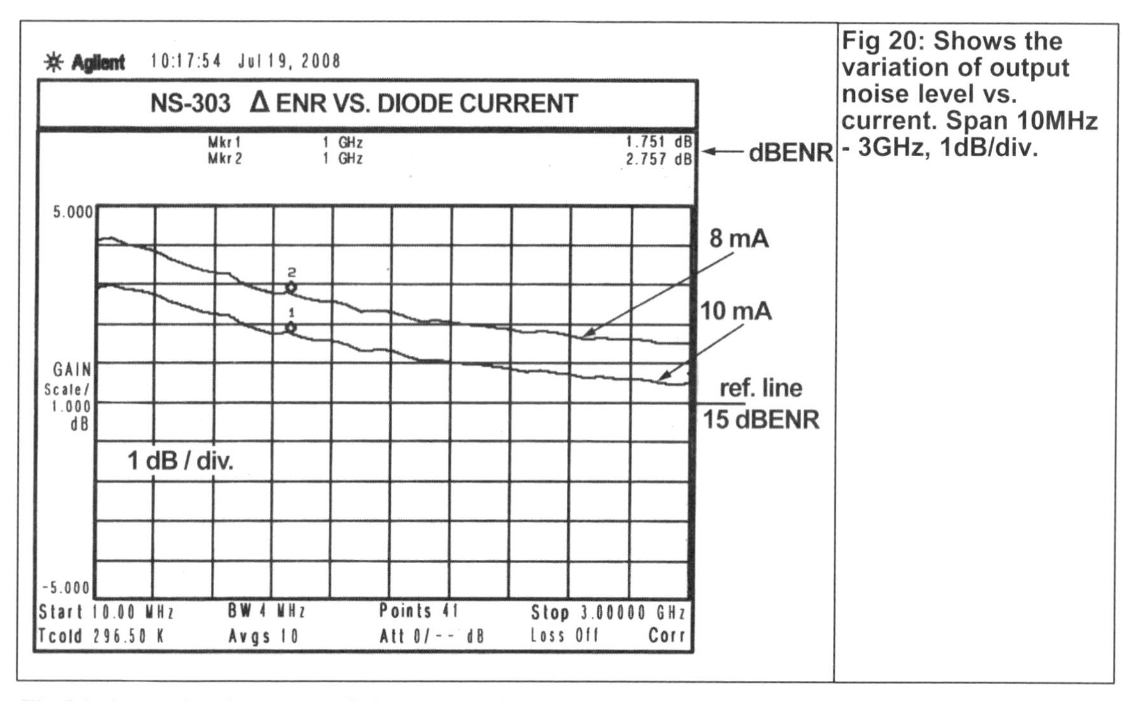

Fig 20: Shows the variation of output noise level vs. current. Span 10MHz - 3GHz, 1dB/div.

Fig 21 shows the decrease of about 1dB of ENR level with 10mA instead of 8mA maintaining the same shape in the diagram.

During the calibration it is possible to play with the current to "tune" the ENR level, if you can loose 1dB of ENR level, you will have a more extended frequency range which is exactly what

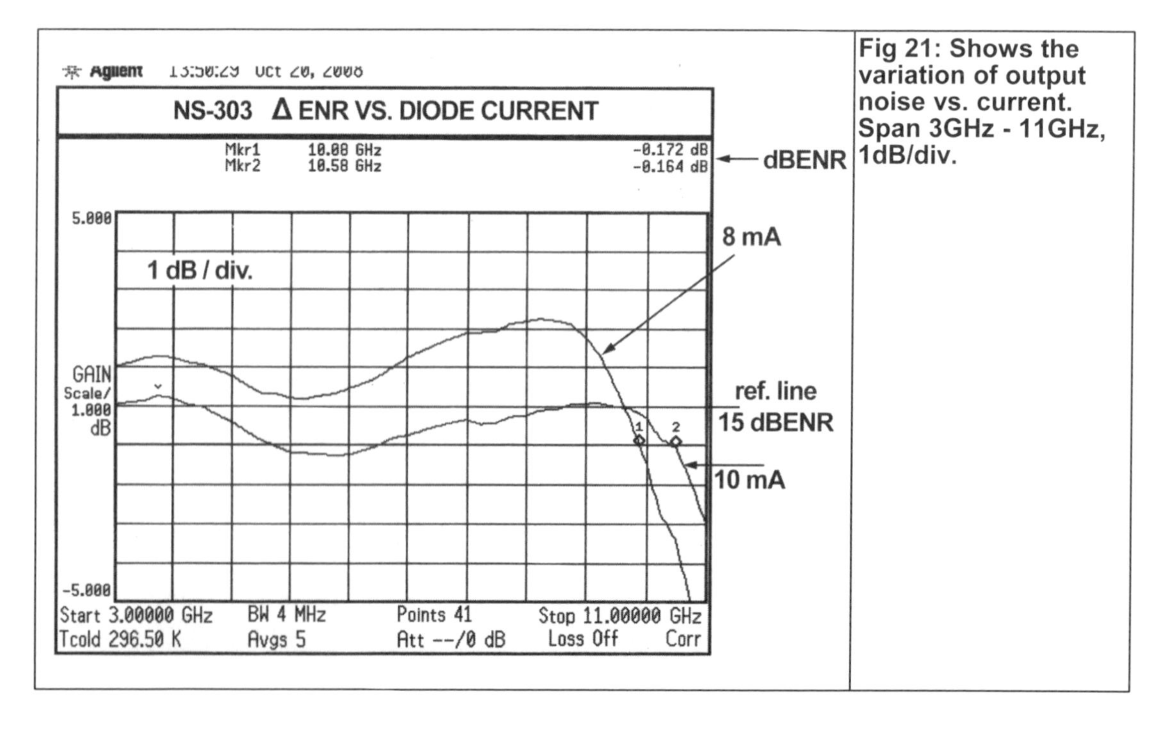

Fig 21: Shows the variation of output noise vs. current. Span 3GHz - 11GHz, 1dB/div.

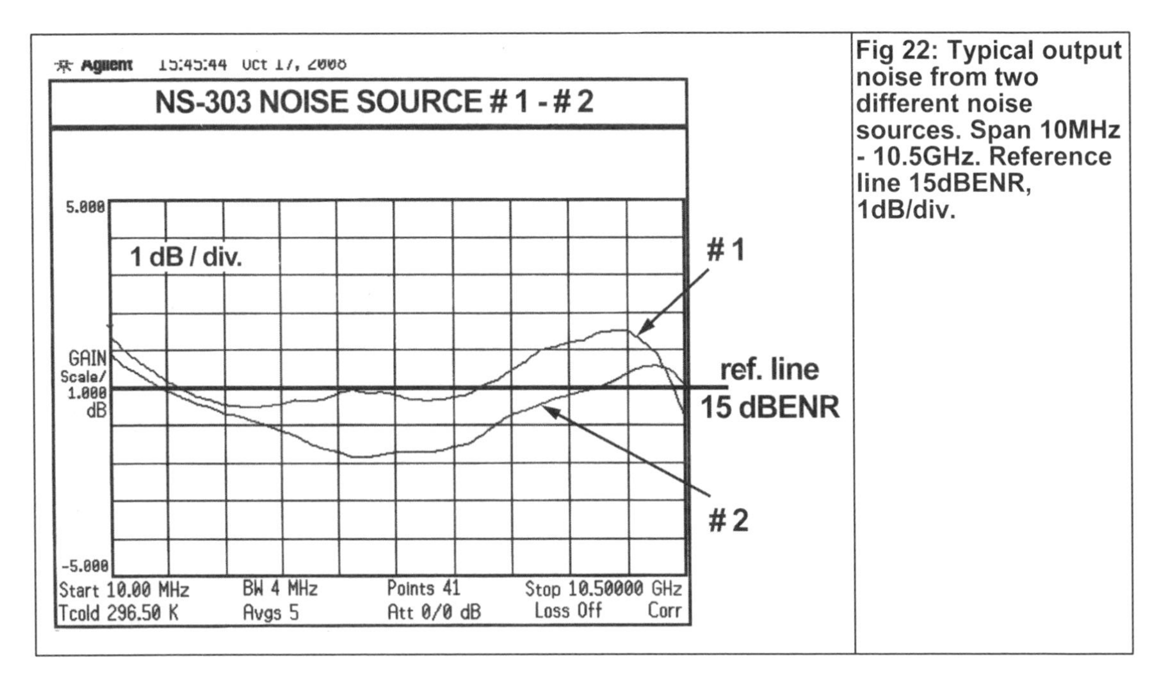

Fig 22: Typical output noise from two different noise sources. Span 10MHz - 10.5GHz. Reference line 15dBENR, 1dB/div.

is needed to reach the 3cm amateur radio frequency band (10.4GHz).

The bias current can be measured easily directly on the BNC input connector with +28Vdc from a normal power supply; the input current is more or less the same current through the noise diode.

Test results

I tested 20 pieces of the noise sources generator and they all gave nearly the same results, the measurement in Fig 22 refer to the use of a 6dB internal attenuator plus a 8dB external attenuator (MaCom or Narda DC -18 GHz).

A typical output noise level can be 15dBENR +/-1.5dB or 15dBENR +/-2dB or 15dBENR +1/-2dB, a ripple of +/-1.5dB or +/-2dB are normal values.

The output return loss depends mainly on the external attenuator; I measured a 30dB return loss up to 5GHz, 28dB up to 8GHz and 25 to 28dB at 10GHz.

We have to consider that each 1dB more of external attenuation will improve the output return loss by 2dB, so if you can use, for instance, an attenuator of 17/18dB you will reach a very good return loss (>30/35dB) with an output noise around 5dBENR.

Calibration

Unfortunately the calibration of a noise source is not an easy thing to do.

We know very well that RF signal generators have an output level precision of typically +/-1/1.5dB and this doesn't worry us, we also know that our power meter can reach +/-0.5dB precision or even better. We need a very high precision for a noise generator used with a noise figure meter. For the classic noise source 346A, B and C, Agilent gives ENR uncertainty of +/-0.2dB max. (< 0.01dB/°C) and 0.15dB max. The new N4000 series are used for the new noise figure analyser N8975A.

In my lab I used the new noise figure analyser N8975A with the precision noise source N4001A so I can guarantee a typical precision of +/- 0.1dB up to 3GHz and 0.15dB up to 10GHz.

It means that the calibration must be done with a good reference noise source, it can be a calibrated noise source compared with the one you have built with a low noise preamplifier and a typical noise figure meter.

Example: you have a low noise amplifier with 0.6dBNF and your calibrated noise source indicates a 15.35dB of ENR, now you can change the noise source to the one you have built and for instance you measure 0.75dBNF, it means that your noise source has 15.35 + (0.75dB - 0.6dB) = 15.50dBENR.

Other application

As I described in the previous article [3] that the noise source can be used as a broadband noise generator combined with a spectrum analyser like a "tracking generator" for scalar applications.

This is not a true tracking generator because it works in a different way (read my previous article [3]). The problem here is to reach 3 decades of frequency range, 10MHz to 10GHz, with a flat amplifier of at least 50dB.

Today some MMICs are available that can do this work like ERA1, ERA2, MGA86576 etc..., the problems can be to reach a flat amplification and to avoid self oscillations with such high amplification.

This device can be very interesting because it can be a useful tool to use with any kind of obsolete spectrum analyser to tune filters, to measure the return loss etc... up to 10GHz.

For more information regarding noise source diodes see [7]

Noise factor measurement with older spectrum analysers [8]

The theory and practice of noise factor measurement plus HOW and WHY older measuring instruments can be used is described in detail in this article.

If you are developing low noise RF amplifiers, the control of noise factor is part of the project. All modern microwave CAD programs have a facility to simulate this but everyone wants to test the practical results.

There are marvellous measuring instruments but the price is a barrier for hobby use. I remember some articles in books and sample applications from the Internet that describe using a spectrum analyser for this task. The question: is an old HP140 with an HP8555 module suitable? What is missing for this measuring task, what must be developed? How do we proceed with these measurements and what theoretical knowledge is necessary? It is also necessary to know what to do, in which order and why it must be done. This article gives the answers to these questions and shows a practical solution.

A short course on the phenomenon of "noise"

"Noise" - where does it come from?

That can be answered quickly and precisely: in each electrical resistance where a current flows and electrons move. As soon as heat comes into play (that is always the case above absolute zero), electrons have an independent existence. They move ever more jerkily and not straight from minus to plus. They collide, rebound, are hurled forward or off to the side.... This makes the current vary irregularly by small amounts due to the influence of heat. This effect is called Thermal Noise. Even if no outside voltage is applied these independent movements of the charge carriers, due to heat, develop a small open circuit voltage V_{noise}. It can be computed as follows:

$$V_{noise} = \sqrt{\frac{4hfBR}{e^{\frac{hf}{kT}} - 1}} \qquad (1)$$

Where:

h = Planck's constant

k = Boltzmann's constant = 1.38 x 10^{-23} J/Kelvin

T = absolute temperature in Kelvin

B = bandwidth in Hz

f = centre frequency of the band in Hz

R = resistance value in ohms

That seems terribly complicated and not practical, but the following can be used without problem to at least 100GHz and temperatures down to 100°K:

$$V_{noise} = \sqrt{4kTBR} \qquad (2)$$

Changing this formula around it suddenly looks much simpler:

$$\frac{\left(\frac{V_{noise}}{2}\right)^2}{R} = kTB \qquad (3)$$

- It is a simple indication of power! Each resistance, independent of its resistance value, produces an "Available Noise Power" proportional to kTB.
- The resistance should be regarded as voltage source consisting of V_{noise} and a noise free *internal resistance* R to generate the noise power. A noise free load resistance with the value R is connected to this source. Because voltage across the load resistance is then half V_{noise} from formula (2) the load resistance receives the Available Noise power kTB.
- This noise power increases linearly with the absolute temperature of the circuit and the voltage with the square root of the power. The spectral power density (power per Hz) is independent of the frequency; this is called "white noise".

Importantly:

Receiver systems are nearly always specified in terms of "power levels" instead of voltages. These are logarithmic measurements so that gain can be calculated by adding levels instead of multiplication. The most familiar unit is "dBm" that is not a voltage but a power rating in relation to the system reference resistance.

P_0 = 1 milliwatt at the system resistance

Thus:

Power level = 10log (power value/1mW) in dBm (4)

If the noise power "kTB" is considered more exactly, an interesting simplification can be introduced:

kTB = (kT) × B = (Noise power density) × (Bandwidth)

The noise power density "kT" represents the power in every Hz; this must be multiplied by the bandwidth in order to calculate the noise power produced. Converting this to a level calculation the following formula should be committed to memory:

Each resistance produces at ambient temperature (T_0 = 290°K) an available noise level and thus an available noise power density of :

-174dBm per Hz of bandwidth.

For bandwidths larger than 1Hz:

Available noise level in dBm = -174dBm + 10log (bandwidth in Hz)

As a small example:

The no load noise voltage V_{noise} at the terminals of a 50Ω resistor for a bandwidth of 100kHz can be calculated:

Noise level with matching = -174dBm + 10log (100000) = -174dBm + 50dB = -124dBm

That results in an available noise power of :

$$P = 1mW \cdot 10^{\frac{-124}{10}} = 1mW \cdot 10^{-12.4} = 4 \cdot 10^{-16} W$$

That will produce voltage across the 50Ω resistor of:

$$V_{noise} = \sqrt{P \cdot R} = \sqrt{50\Omega \cdot 4 \cdot 10^{-16} W} = 141nV$$

So the open circuit voltage V_{noise} is twice this value, thus 282nV.

Other references:

On the Internet there is a lot of free literature on this topic. The chapter "Noise" in the standard work on microwaves by David Pozar [9] seems to be the model reference. Unfortunately this book is very expensive and therefore a better reference is probably the most important HP Agilent application note [10]. It also contains all basic information. In addition the application notes [11] and [12] from the same company are worth downloading. The application note from MAXIM [13] on the topic of noise measurement is also worth having on your computer.

Further sources of noise

Each active component (valve, bipolar transistor, barrier layer FET, MOSFET, HEMT etc.) produces two additional types of noise:

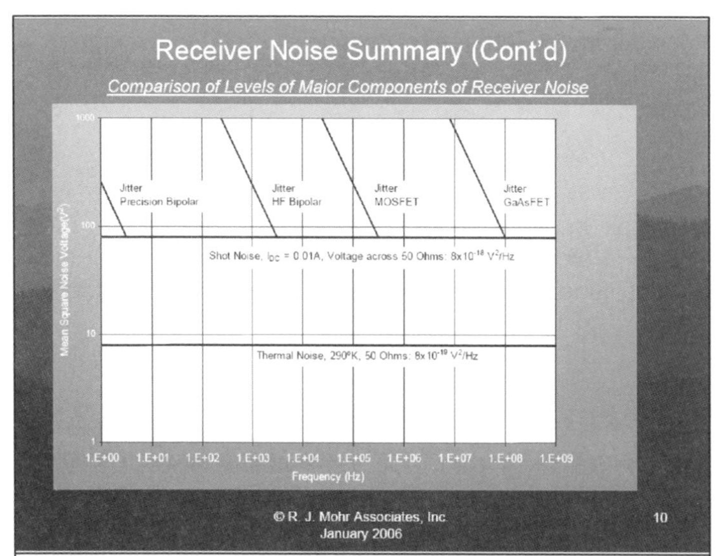

Fig 23: This diagram from [14] shows the three different noise generating mechanisms mentioned in the text.

- **Shot Noise** produced in valve diodes and pn transitions by distortions of the current flow when crossing the potential differences. It produces wide band, white noise.
- **Flicker Noise** or "1/f – Noise" (sparkling noise) results from defects in the crystal structure caused by impurities. They lead to short pulse type fluctuations in the current flow that produce a spectrum whose power density decreases with rising frequency. There exists a corner frequency and it is interesting to see how this differs between different active components. A good impression of this can be found in Fig 23. The reference material in [14] should be obtained from the Internet, it contains a precise but compact introduction to the subject.

Before semiconductors, gas discharge tubes were used as sources of noise, they produce a very wide band plasma noise.

The following section shows how the different kinds of noise in a circuit can be considered and how they can be summarised by only one parameter.

Noise temperature and noise factor for a two-port system

When components of a communications system are specified they usually have the same system resistance (normally 50Ω, but for radio, television and video systems, 75Ω) and signal quality information required for correct transmission. This is often expressed by the SINAD value (Signal to Noise and Distortion). The maximum signal level is limited by over modulation, distortions, intermodulation etc. The noise "floor" (in addition, other disturbances e.g. cross modulation from other channels) gives the minimum signal level in the system. Therefore each component must be specified by its S parameters a well as its noise behaviour.

Regarding the noise, the following data is possible and usual (unfortunately not all authors use the same terminology)

- The equivalent noise temperature of a component expresses the self-noise of the component by an additional temperature rise for the source resistance. The component itself is thought of as noise free.
- The Noise Figure "NF" indicates, in dB, how much worse the signal-to-noise ratio becomes after the signal has passed through the component (this is a

measure of how the noise from the internal resistance is added to the signal).

- The Noise Factor "F" is like the Noise Figure as above, BUT not in dB, but as a simple gain relationship.

The noise temperature

Consider a two-port component, for example an amplifier for a 50Ω system. The signal source must have an internal resistance of 50Ω in order to give a power match with the 50Ω input impedance. If the input signal has an amplitude of zero, therefore only noise is present at the output.

Now we proceed in two steps:

- First consideration: the amplifier is noise free; it sees only the internal resistance of the source. This has a temperature of T_0 = 290°Kelvin (standardised ambient temperature world-wide) therefore the noise signal supplied to the matched amplifier input is:

$$P_{N\_IN1} = kT_0B$$

- Second consideration: the source resistance is cooled down to zero degrees Kelvin and therefore produces no noise. The input signal is still zero. Everything that is measured at the amplifier output is noise that comes from the amplifier itself. Now consider that the amplifier circuit with all its components is noise free and the measurable noise comes just from an additional heating of the source resistance. The apparent temperature necessary for this is called "equivalent noise temperature T_e" and the associated power can be calculated by:

$$P_{N\_IN2} = kT_eB$$

The amplified sum of these two noise powers can be measured at the output (this is permitted, because the two noise parts are not correlated and may be added). The formula for the power amplification of such a component is the linear relationship of the two powers, thus the "linear power Gain" is:

$$G_{P\_LINEAR} = \frac{P_{OUT}}{P_{IN}}$$

This must be calculated if only the dB power amplification is given:

$$G_{P\_LINEAR} = 10^{\frac{G_{P\_dB}}{10}} \qquad (5)$$

Thus the entire available noise power at the component output is:

$$P_{N\_OUT} = G_{P\_LINEAR} \cdot (P_{N\_IN1} + P_{N\_IN2})$$

$$= G_{P\_LINEAR} \cdot (kT_0 B + kT_e B)$$
$$= G_{P\_LINEAR} \cdot k \cdot B(T_0 + T_e)$$

This shows a very beautifully thing concerning the additional temperature rise of the source resistance. If the ambient temperature T_0 is rearranged it gives:

$$P_{N\_OUT} = G_{P\_LINEAR} \cdot kT_0 B \cdot \left(1 + \frac{T_e}{T_0}\right) \qquad (6)$$

If this expression is divided by the power amplification G_{P_LINEAR}, it gives the effective noise power at the amplifier input :

$$P_{N\_IN\_TOTAL} = \frac{P_{N\_OUT}}{G_{P\_LINEAR}} = kT_0 \cdot \left(1 + \frac{T_e}{T_0}\right) \qquad (7)$$

It becomes clear that the 1 in the brackets represents the ambient temperature noise from the source resistance. The other term indicates how much noise comes from the amplifier and must be added (e.g. if the equivalent noise temperature of T_e is the same as the ambient temperature T_0, the input noise performance will be doubled). Thus the bracketed term is the noise factor “F”.

The noise factor F and the Noise Figure NF

The noise factor is:

$$F = \left(1 + \frac{T_e}{T_0}\right) \qquad (8)$$

If the equivalent noise temperature of a component is required and the noise factor F is known, the conversion of this formula is child’s play:

$$T_e = T_0 \cdot (F - 1) \qquad (9)$$

Therefore a *noise factor* F = 1 results in an equivalent noise temperature of zero Kelvin and designates a completely noise free component (only the amplified noise of the source resistance appears at the output).

It is very often practical to express the noise factor in dBs. This “Noise Figure NF in dBs” is determined as follows:

$$NF = 10\log(F) = 10 \cdot \log\left(1 + \frac{T_e}{T_0}\right) \qquad (10)$$

Unfortunately there are considerable discrepancies for this in literature, but we want to adhere to the English habit in this article:

F = noise factor *(linear relationship)*

Noise Figure NF = 10 log (F) in dBs *(logarithmic relationship)*

A practical note for a passive component:

- The Noise Figure NF of a passive component corresponds its attenuation in dBs

A 20dB attenuator has a Noise Figure of 20dB this is easy to explain. Assume that an antenna is directly connected to the input of a receiver. As well as the information signal the receiver input receives the self-noise signal from the internal resistance of the antenna. If a 20dB attenuator is connected between the antenna and the receiver the two antenna signals (noise and information signal) will be weakened by 20dB. The receiver sees the internal resistance of the 50Ω attenuator that creates noise from the ambient temperature i.e. kT_0B. The noise level at the attenuator output in relation to the divided down portion suddenly increases by 20dB and of the Noise Figure increases by 20dB according to the definition!

Here is a summary of the most important parts so far:

The Noise Figure (NF) indicates by how many dBs the signal-to-noise is worsened when goes through the component or system (note: the source supplies not only the desired information signal, but also noise that comes from its internal resistance).

Converting the dB value of the Noise Figure into a linear relationship gives the Noise Factor (F).

Series connection of components

In practice it is usual to deal with a whole chain of components or building blocks where each one strengthens and weakens differently. How can the noise behaviour of the complete system be calculated? It can be done in the following way:

Each individual component of the chain has its own self-noise (usually the Noise Figure NF in dB) and its power amplification (power gain in dB) described by the S parameters S21. A signal voltage with an information signal level S_{IN} and an internal resistance, whose value agrees with the system resistance, are applied to the input of the chain. Now two things are of interest:

- How large is the information signal level at the output that gives the overall gain of the arrangement?
- What are the S/N values at the input and output? By what factor and dB value has the S/N worsened at the output as opposed to the input?

The first question can be answered quickly because it only requires the addition of the gain values (in dB) of all building blocks involved, considering the sign, in order to calculate the overall gain. The sum of input level in dBm and overall gain in dB results in the output level in dBm.

The second question is considerably more complex, because here we MUST NOT use the logarithmic values.

The correct result is only obtained using the linear values (Noise Factor F in place of Noise Figure NF in dB and power amplification G_{P_LINEAR} in place of power gain in dB). This gives:

$$F_{TOTAL} = F_1 + \frac{F_2 - 1}{G_{P\_LINEAR\_1}} + \frac{F_3 - 1}{G_{P\_LINEAR\_1} \cdot G_{P\_LINEAR\_2}} + \frac{F_4 - 1}{G_{P\_LINEAR\_1} \cdot G_{P\_LINEAR\_2} \cdot G_{P\_LINEAR\_3}} \text{.........} \quad (11)$$

This shows the well-known fact that only the noise of the first stage is all included in the total noise factor. Notice the second stage contribution is less because of the higher gain of the first stage. The stages following on have insignificant noise contribution with sufficiently high gains.

Please remember: the entire noise is calculated by this formula as all being in the overheated source resistance while the entire system following is accepted as noise free.

Dealing with a noisy signal

This section deals with a practical question that is either treated very badly in textbooks or not at all:

How does the signal-to-noise ratio of a signal at the output of a component (with a power gain G_1 *and* a Noise Figure NF_1) worsen if the input is a noisy signal? (This means that the noise floor is already above the theoretical minimum of kT_0B).

This situation is always true for a building block WITHIN a system chain e.g. the signal-to-noise change in an if amplifier due to a signal from the preceding mixer or, analysing the second or third stage of a 3 stage satellite LNB before the mixer.

From the previous section, the effect of the self-noise of a component is smaller if the preceding component has enough gain to overcome to noise but this must be properly proven.

Looking at Fig 24 in detail it can be seen that the component concerned receives several input signals:

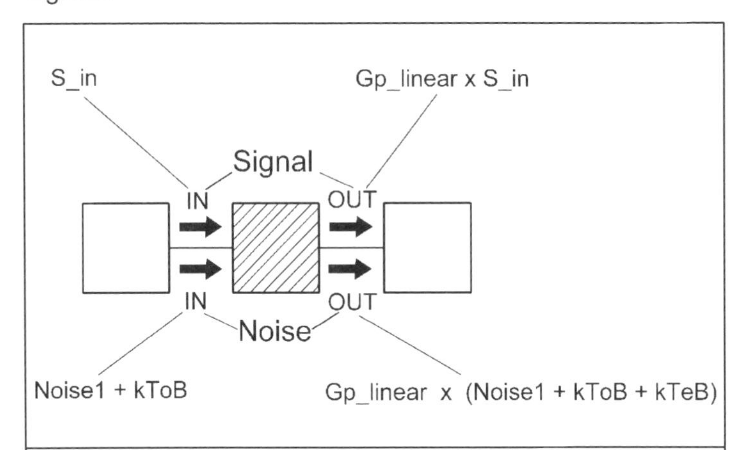

Fig 24: The signal and noise can be analysed seperately.

- The information signal S_{IN} including the noise signal $Noise_1$, supplied by the preceding stage
- The noise power kT_0B from the source resistance (internal resistance of the preceding stage)

So the output signal can be divided into 4 signals:

- The information signal amplified by G_{P_LINEAR}
- Three noise powers that can be added since they are not correlated. They are: the noise from the previous stage amplified by the gain (G_{P_LINEAR} x $Noise_1$), The amplified noise from the source resistance (G_{P_LINEAR} x kT_0B) and the amplified self-noise referred to the input of the component (G_{P_LINEAR} x kT_eB).

The signal-to-noise ratio at the output can be written:

$$\left(\frac{S}{N}\right)_{OUT} = \frac{G_{P\_LINEAR} \cdot S_{IN}}{G_{P\_LINEAR} \cdot \left(Noise_1 + kT_0B + kT_eB\right)} = \frac{S_{IN}}{Noise_1 + kT_0B + kT_eB} \quad (12)$$

If the expression in the denominator $Noise_1 + kT_0B$ is removed (the portion that is supplied to the component input), the answer required is achieved:

$$\left(\frac{S}{N}\right)_{OUT} = \frac{S_{IN}}{\left(Noise_1 + kT_0B\right) \cdot \left(1 + \dfrac{kT_eB}{Noise_1 + kT_0B}\right)} = \frac{\left(\dfrac{S}{N}\right)_{IN}}{\left(1 + \dfrac{kT_eB}{Noise_1 + kT_0B}\right)} \quad (13)$$

Changing the formula around gives:

$$\left(1 + \frac{kT_eB}{Noise_1 + kT_0B}\right) = \frac{\left(\dfrac{S}{N}\right)_{IN}}{\left(\dfrac{S}{N}\right)_{OUT}}$$

This is a familiar expression that represents the linear form of the statement already made:

> *The signal-to-noise ratio (in dB) at the output of the component decreased by the Noise Figure NF (in dB).*

Calling the term in the left bracketed the current noise factor F_{actual} gives;

$$F_{actual} = 1 + \frac{KT_eB}{Noise_1 + kT_0B} \quad (14)$$

This formula can be validated easily:

- The supplied noise $Noise_1$ is zero. This gives the familiar relationship for the noise factor F:

$$F_{actual} = 1 + \frac{T_e}{T_0}$$

- The supplied noise $Noise_1$ is much larger than the noise produced in the stage and the noise of the effective source resistance for this stage at the input. Then the value of the second term in formula (14) is zero therefore:

$$F_{actual} = 1$$

That is also logical, because a high noise level in front of the stage means that the small self-noise does not need to be considered.

The unfortunate fact remains that the dB values of gain cannot be used they must be converted into linear gains.

That was the theory; now the practice follows!

Determining of the noise figure F

In principle this can be carried out using three different methods with appropriate variations:

First method:

Some years ago there was a noise-measuring instrument available, The noise power at the output was adjustable and the scale of the instrument was calibrated in kT_0 or dB. This transmitter was connected to the input of the test specimen and a wattmeter was connected to the output. Now the transmitter was adjusted until the output noise power of the test specimen rose by 3dB (DOUBLED POWER). Therefore the input noise power was equal to the noise produced in the test specimen, this could be accurately read from the transmitter. Simply, but effectively!

The only problem if you find one of these at a flea market is that they use a special valve noise diode. These are rareparts and age very quickly. For a problem free "gem" look for similar equipment with a more modern semiconductor noise generators.

Second method (Gain method):

Measure the noise power at the output of the test specimen with a low noise spectrum analyser and the input terminated with the system impedance (usually 50Ω). If the test specimen has sufficient high gain, the self-noise of the analyser can be neglected. For the power output:

$$N_{OUT} = (Gain) \times (kT_0 Bandwidth) \times F \quad \text{in Watts}$$

The advantage of this method is only apparent if the power level is changed to dBm. The formula is then:

$$N_{OUT}(dBm) = (Gain) + \frac{-174 dBm}{Hz} + 10 \cdot \log(Bandwidth) + NF$$

To make the measurement the gain of the test specimen in dB must be known. The spectrum analyser displays the power output in dBm on the screen and by using the controls the power can be read. The remainder is now a simple dB calculation:

$$NF = N_{OUT} - (Gain) - 174\, dBm - 10 \log(Bandwidth)$$

Remember that this method assumes that the self-noise of the spectrum analyser can be neglected. Therefore it only works correctly for test specimens with high self-noise or very high gain

Third method (Y factor method):

The internal resistance is connected to the test specimen input at two different temperatures, T_{hot} and T_{cold} and measure the noise power at the output in both cases.

Working with two temperatures is no joke; using liquid helium, for the low temperature and boiling water for the high temperature! Receiver measurements can be made with the antenna pointed towards the cold sky and then towards the warm earth. Alternative, some decades ago, special thermionic diodes or gas discharge tubes were used.

Modern noise figure measuring instruments use this principle, however special semiconductor noise diodes, usually avalanche diodes, are used. These devices produce ambient temperature noise when switched off. When they are switched on they produce a known noise power corresponding to a hot resistor. This cold to warm difference is indicated as the ENR value in dB (Excess Noise Ratio).

The measuring procedure is quite simple. With T_{HOT} and T_{COLD} the following formulae are used for the output noise power:

$$N_{OUT\_COLD} = k \times Gain \times Bandwidth \times \left(T_{COLD} + T_{equivalent}\right)$$

$$N_{OUT\_HOT} = k \times Gain \times Bandwidth \times \left(T_{HOT} + T_{equivalent}\right)$$

These are two equations for two unknown values i.e. for the equivalents noise temperature $T_{equivalent}$ of the test specimen and for the Gain Bandwidth Product (which is of less interests). Mathematics and the microprocessor in a modern Noise Figure meter have no problems in calculating the equivalent noise temperature and then the noise figure.

$$F = \left(1 + \frac{T_{equivalent}}{T_0}\right)$$

A further calculation step gives the Noise Figure in dB.

$$NF = 10 \cdot \log\left(F\right)$$

Noise measurements using your own older spectrum analyser

Measurement principle

Unfortunately the gain method cannot be used with older spectrum analysers because of their high self-noise and limited sensitivity. Therefore a different procedure must be used:

- A low-noise wideband preamplifier is used in front of the spectrum analyser. A maximum Noise Figure of 3 to 4dB as well as a gain of at least 40 - 60dB is required. The output resistance must correspond as exactly as possible to the system resistance. S11 and S22 should not be larger than -10 to –15dB in order to exclude errors due to reflections and miss matches. Using this amplifier the noise level arriving at the analyser input is increased so that the analyser self-noise no longer plays a role.
- The input of the preamplifier is terminated with a 50Ω resistor and the noise signal output is set to the centre of the spectrum analyser screen. Decrease the video bandwidth of the spectrum analyser until the noise floor is just a straight line this is now:

  THE LINE THAT REPRESENTS THE NOISE POWER kT_0B OF THE RESISTANCE

This not only eliminates the self-noise of the additional preamplifier and the spectrum analyser but also simulates (as required) a noise free source resistance.

- The remainder is very simple and clarified by an example:

If the 50Ω termination on the input is replaced with a test preamplifier with 10dB gain and the 50Ω termination connected to the input of this test preamplifier. The screen display will increase by exactly 10dB if the test preamplifier is noise free. If the display increases for example by 11.5dB the output S/N had decreased by 11.5dB – 10dB = 1.5dB which is a Noise Figure of 1.5dB.

The fascinating thing is that the frequency response of the additional preamplifier and the analyser play no role since all fluctuations of the gain are balanced by the calibration procedure but the minimum gain present must overcome the self-noise of the spectrum analyser.

This method is only successful if:

- The gain of the test specimen (in dB) is very very accurately known. It also relies the accuracy of the spectrum analysers logarithmic amplifier.
- There are extremely high demands placed on the design of the preamplifier (absolutely oscillation free wide band with over 40dB but up to 60dB gain and a good match at the input and output). It is a headache to design but with experience of designing microwave amplifiers it is possible.

The preamplifier

Starting the development

MMICs with at least 20dB gain and an upper cut-off frequency of at least 2.5GHz are used in the amplifier stages. This means that different interesting ranges can be covered like “government bands”, “Meteosat” or the 13cm amateur band but not the lower amateur radio frequencies. The amplifier must absolutely stable and oscillation free, which can surely be achieved using simulations.

Three such building blocks can be connected in series in a long enclosure to realise the minimum gain of 60dB required. Extensive precautions must be employed to suppress the inevitable self-oscillation.

Finally the Agilent ABA-52563 broadband amplifier was chosen, the data sheet begins:

Agilent ABA-52563: 3.5GHz Broadband silicon RFIC Amplifier

Feature:

- Operating frequency: DC ~ 3.5 GHz
- 21.5dB gain
- VSWR < 2.0 throughout operating frequency
- 9.8dBm outputs P1dB

- 3.3dB noise figure
- Unconditionally stable
- Single 5V supply (ID = 35mA)

Development of the single stage amplifier prototype

The S parameters for the ABA-52563 are required, but unfortunately they are only printed as table in the data sheet. Thus they must be copied from the table, deleting unnecessary data, into the required S2P form as "ABA52563. s2p " in the project directory, this is shown in Table 2.

The result of the first PUFF simulation, only using the S parameters, is shown in Fig 25 (it is clearer and faster with PUFF rather than with the Ansoft designer student version). The result looks good in the intended frequency range from 500 to 2500MHz; S21 is approximately constant at 21db, S11 and S22 at -15dB.

Fig 26 shows the practical circuit diagram on the left. It has extensive decoupling of the supply to prevent noise, particularly low frequencies on the supply line (e.g. self-noise of the power

Table 2: The S parameters for the MMIC ABA 52563 from Agilent.

```
! ABA-52563 S PARAMETERS
# ghz s ma r 50
```

0.05	0.01	146.6	12.10	-2.6	0.03	0.3	0.15	-2.4
0.10	0.01	134.0	12.11	-4.8	0.03	-0.3	0.15	-5.1
0.20	0.01	-40.6	12.16	-9.6	0.03	0.1	0.15	-9.6
0.30	0.01	-53.2	12.19	-14.5	0.03	1.2	0.15	-13.0
0.40	0.02	-56.7	12.19	-19.5	0.03	2.4	0.14	-15.7
0.50	0.03	-141.5	12.26	-24.8	0.03	1.0	0.15	-15.7
0.60	0.03	-128.1	12.24	-29.8	0.03	3.1	0.15	-17.6
0.70	0.04	-127.5	12.21	-34.9	0.03	4.3	0.15	-20.3
0.80	0.04	-126.7	12.18	-39.8	0.03	6.1	0.15	-22.5
0.90	0.05	-123.9	12.16	-44.7	0.03	7.4	0.15	-24.2
1.00	0.05	-125.0	12.13	-49.7	0.03	11.7	0.15	-26.4
1.20	0.05	-123.4	12.10	-59.6	0.03	10.8	0.15	-29.4
1.40	0.06	-127.4	12.05	-69.4	0.03	12.4	0.15	-32.4
1.60	0.06	-133.8	12.04	-79.6	0.03	13.0	0.15	-35.3
1.80	0.06	-136.7	12.00	-89.8	0.04	14.7	0.15	-37.8
2.00	0.07	-142.5	11.94	-100.4	0.03	14.3	0.15	-38.3
2.20	0.07	-143.9	11.87	-111.2	0.04	16.7	0.15	-37.8
2.40	0.08	-146.1	11.75	-121.9	0.04	16.2	0.15	-37.3
2.60	0.09	-148.4	11.56	-133.2	0.04	17.3	0.14	-36.9
2.80	0.09	-149.5	11.33	-144.5	0.04	15.6	0.14	-36.4
3.00	0.10	-152.7	10.95	-156.1	0.04	15.8	0.13	-35.9
3.20	0.10	-158.7	10.51	-167.5	0.04	15.6	0.13	-35.4
3.40	0.11	-163.2	9.97	-478.7	0.04	15.5	0.13	-34.9
3.50	0.11	-167.6	9.67	175.9	0.05	16.0	0.13	-34.6
4.00	0.12	165.9	8.25	150.6	0.05	12.0	0.13	-33.4
4.50	0.16	138.3	6.98	126.3	0.05	12.7	0.14	-37.1
5.00	0.19	122.8	5.71	105.0	0.06	9.5	0.12	-48.4
5.50	0.25	112.3	4.85	86.7	0.07	6.0	0.12	-63.0
6.00	0.30	99.3	4.14	70.4	0.07	1.0	0.11	-83.5

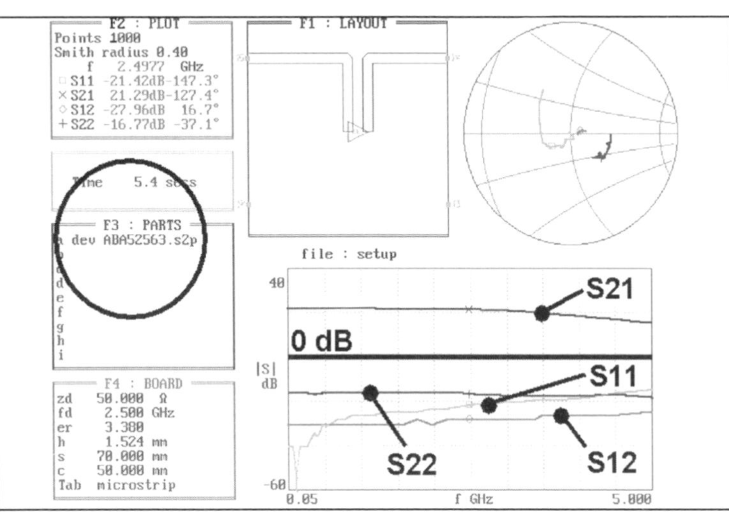

Fig 25: All S parameters of the ABA52563 showing their dependence on the frequency.

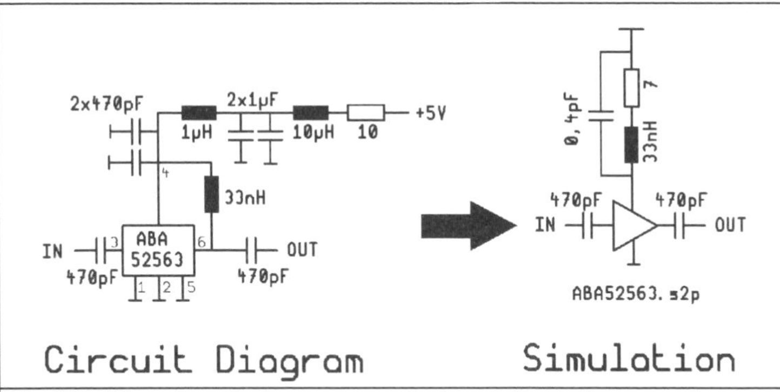

Fig 26: The circuit diagram and simulation diagram for a single stage amplifier.

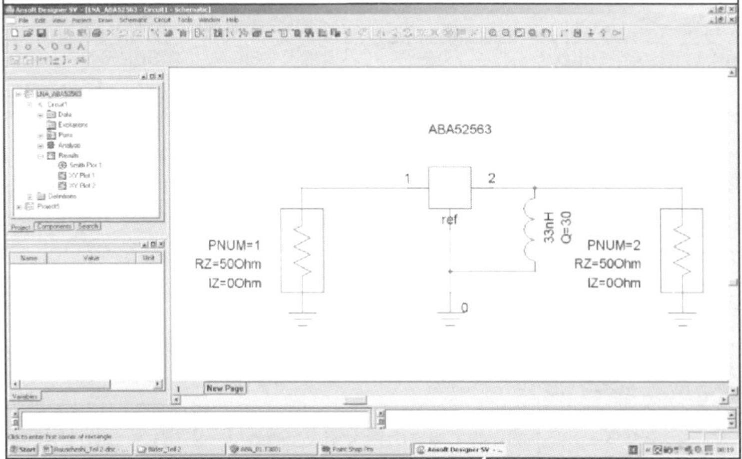

Fig 27: Ansoft designer simulation, the 33nH choke uses the Ansoft model.

supply) from reaching the RF output via the 33nH choke. The right hand side of Fig 26 shows the simulation diagram that the CAD programs (PUFF [15] or AnSOFt designer student version [16]) are content.

The choke was purchased as a finished 0805 SMD component and it should be represented correctly in the simulation with its self-resonant frequency of 1.5GHz and Q of 30 at 1GHz. PUFF requires some preliminary work to generate a replacement equivalent circuit (series resistance = 7Ω, parallel capacitance = 0.4 pF). Ansoft designer student version is much easier to use: there is the construction unit of "chip Inductor" with adjustment possibilities for the individual data and even three different types of representation to select. This is a further reason to only use PUFF as an "RF-pocket calculators" for fast derivation of fundamental solutions and to make the detailed analysis or finished development of the circuit (up to the gain simulation and stability control) with the Asoft designer. Therefore only Ansoft designer was used for further development. Fig 27 shows the editor screen for the simulation of the circuit from Fig 26.

The simulation of the S parameters for the intended frequency range from 500MHz to 2500MHz is shown in Fig

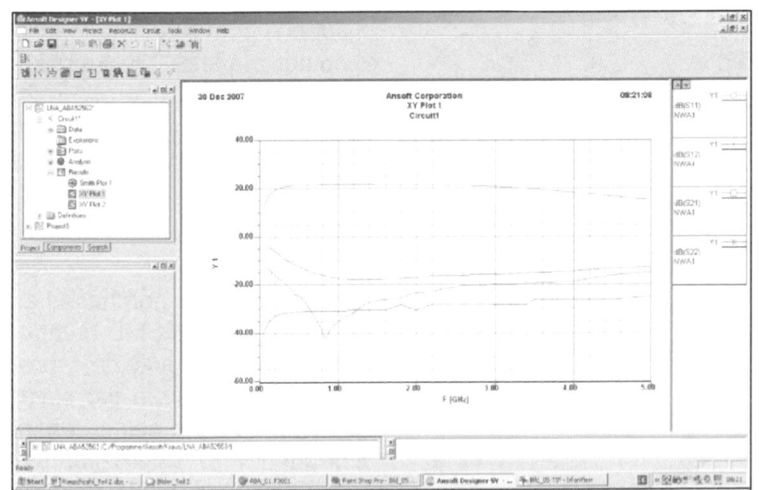

Fig 28: The S parameter simulation of the circuit with a realistic model of the choke shows no surprises.

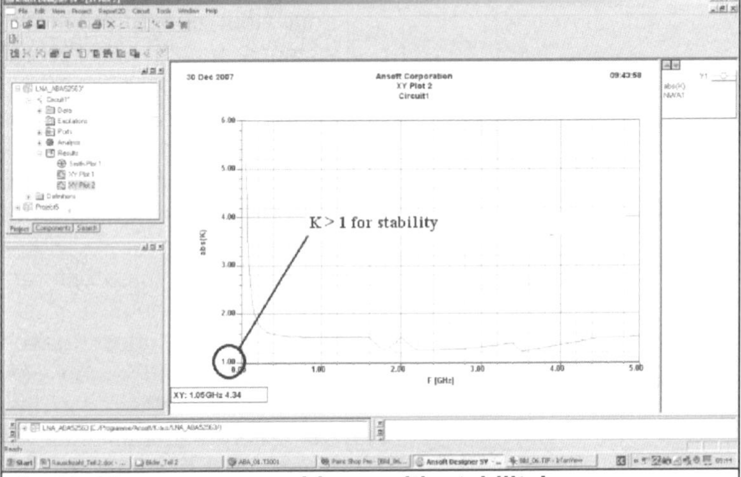

Fig 29: There are no problems with stability!

28. As expected there are no bad surprises, the fall off in gain below 200MHz is inevitable because of the resistance of the choke used.

Following a successful check on the stability of the amplifier (Fig 29) the first attempt to design the PCB was carried out using double-sided 32MIL (0.82mm) thick Rogers RO4003 material. The layout for a 30mm x 50mm PCB is shown in Fig 30. Who is surprised at the apparently unnecessary length of the PCB in relation to the small space required of the circuit? This is one of the precautionary measures to prevent self-oscillation when three amplifiers are connected in series on the same PCB.

There is another topic to consider regarding stability. Those who have already developed circuits using the ABA 2563 will know that the influence of the plated through holes around pins 1, 2 and 5 is extremely critical. If their self-inductance not kept as small as possible there are substantial problems. Fig 31 shows the simulation result for "k" with an inductance only 0.2nH between the earth references of the MMIC (pins 1, 2 and 5) and the ground surface. The simulation circuit is shown in Fig 132. It should be noted that the lower end of chip inductor on the output is connected to the top side of the PCB which is grounded (with 2 high-quality SMD ceramic capacitors joined in parallel) and therefore must be attached to pin 1, 2. The simulation result shows at the very small value of 0.2nH plated-through hole inductance the circuit

Fig 30: The layout of the 30mm x 50mm PCB (lower surface is a continuous ground)

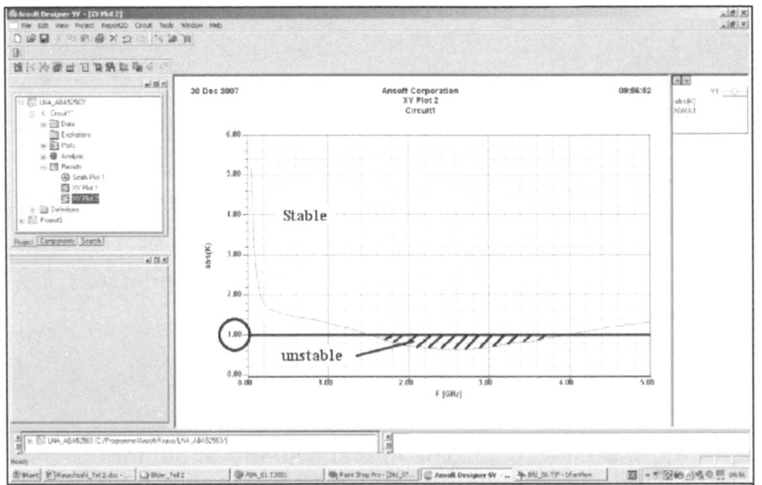

Fig 31: With 0.2nH added for the plated-through hole inductance the circuit becomes useless.

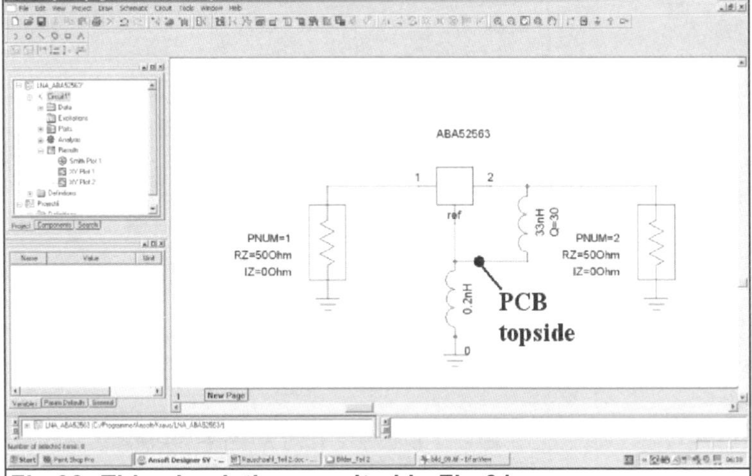

Fig 32: This simulation resulted in Fig 31.

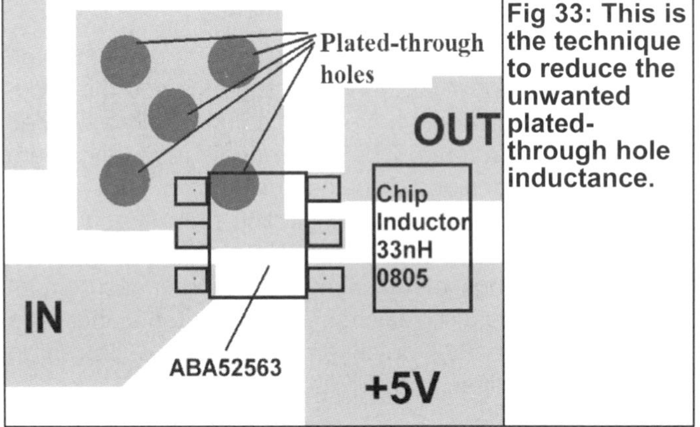

Fig 33: This is the technique to reduce the unwanted plated-through hole inductance.

could not be used because it would oscillate! Fig 23 shows that by enlarging the number of plated-through hole on the PCB layout it is possible to reduce the self oscillation posibility below the danger level. So when placing these holes as close as possible to the pins they are in parallel and the total inductance is reduced dramatically Finally it is just as important to isolate the ground surface on the topside into an input and an output area. Both are connected by a sufficient number of plated-through holes to the continuous ground surface on the bottom side of the PCB.

Fig 34 shows the circuit built into a milled aluminium enclosure, it is easy to locate the positions of the different components of the circuit diagram (please look carefully at the supply voltage which is fed carefully, using a coaxial cable to prevent any stray effects, to the SMB socket on the long side of the enclosure). The results of measurements on the prototype are exciting; these are shown in Figs 35 to 38. It is very pleasing that with the network analyzer (HP8410 with S parameter plug-in HP8746B for 0.5 to 12.4GHz and sweep oscillator HP8690 with HP8699 plug-in for 0.1 to 4GHz) the forecasts can be confirmed so beautifully. The reverse path was also taken; several cases with different plated-through hole inductances were simulated and showed on the basis of the measured S21 value

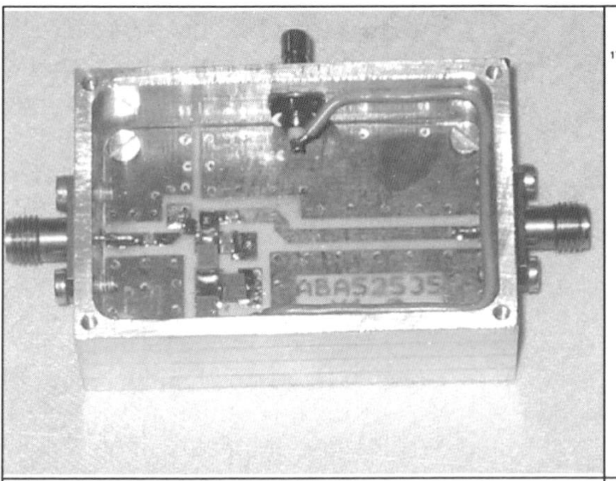

Fig 34: The circuit ready for operation, built into a milled aluminium enclosure. The SMB socket for the voltage supply is on the long side.

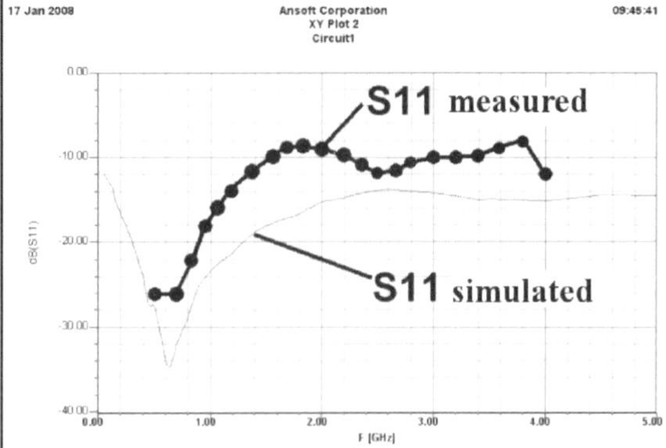

17 Jan 2008
Ansoft Corporation
XY Plot 2
Circuit1
09:45:41
S11 measured
S11 simulated
F [GHz]

Fig 35: The measured and simulated results for reflection S11.

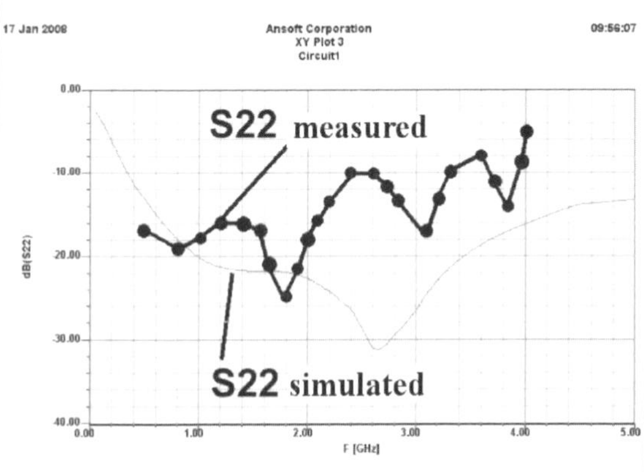

Fig 36: The measured and simulated results for output S22.

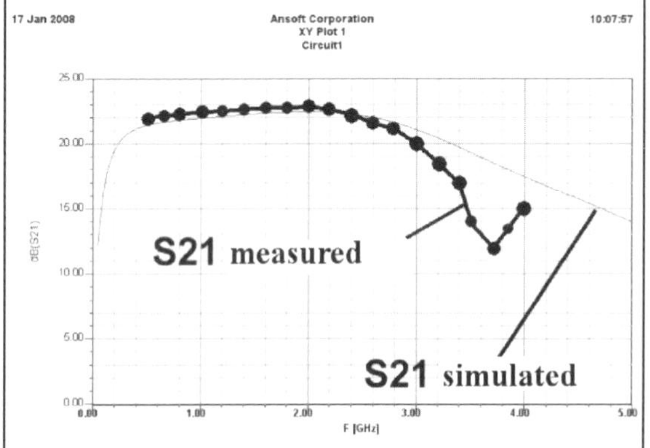

Fig 37: The measured and simulated results for S21.

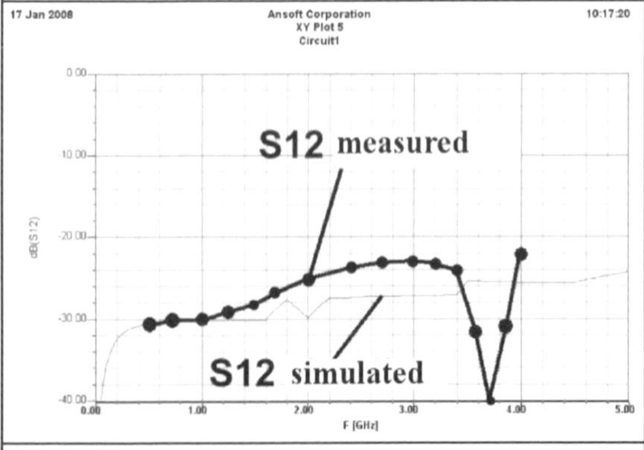

Fig 38: The measured and simulated results for S12.

that a value of 0.07nH gives the best agreement with the measurements.

In order to find the causes for the unexpected response at the high frequency end the simulation would have to be substantially extended and refined. Obvious periodic resonances are happening. Genuine micro strip lines at the input and output, including their effects, have not been considered. Nevertheless; the basic findings of the simulation and measurement agree also the circuit remains stable, therefore the actual development target i.e. the large amplifier, can continue.

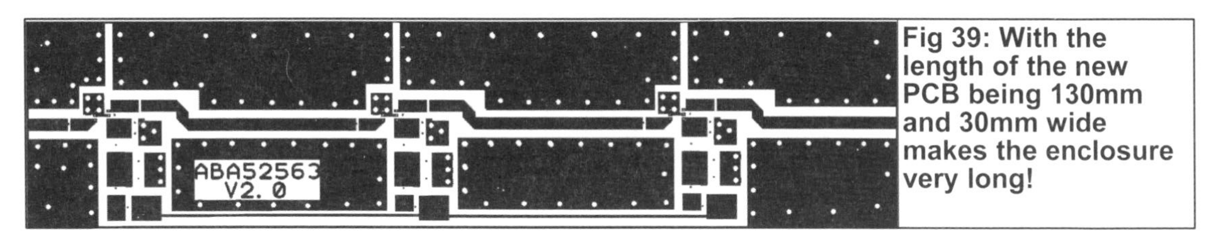

Fig 39: With the length of the new PCB being 130mm and 30mm wide makes the enclosure very long!

The three stage amplifier

The way to the goal

Now it becomes excitingly; how does series connection of three of such circuits in a single aluminium enclosure behave? As a precaution the complete arrangement was simulated in order to guarantee that the interaction of each stage (expressed by its parameter S12) does not become unstable. Because if it were unstable there would not be much sense in further development and an alternative would be sought.

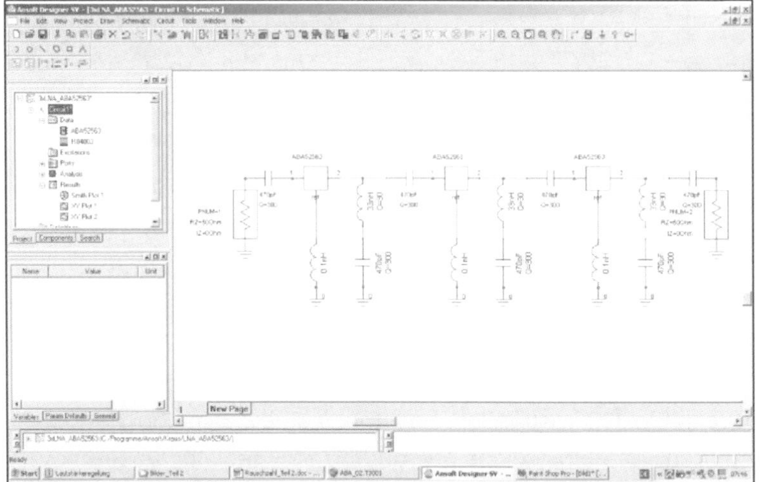

Fig 40: Simulation circuit for the three stage amplifier.

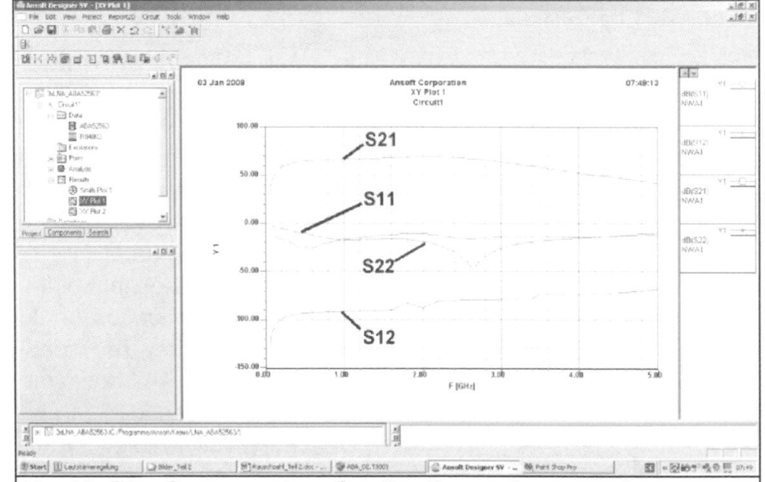

Fig 41: The S parameters for the three stage version simulation are completely normal.

The PCB used to connect the individual stages using 50Ω microstrip is shown in Fig 39. The simulation circuit shown in Fig 40 gives the S parameter simulation shown in Fig 41 and for stability simulation shown in Fig 42. Everything looks good and it now depends on: the coupling factors in the enclosure, the mutual shielding of the individual stages, the correct arrangement of the ground surfaces and their plated-through holes as well as the decoupling of the three stages power supplies whether this large amplifier self-oscillates!

The first prototype was made and screwed into the enclosure (Fig 43). A 50Ω resistor was connected to the SMA input socket with a spectrum analyser connected to the

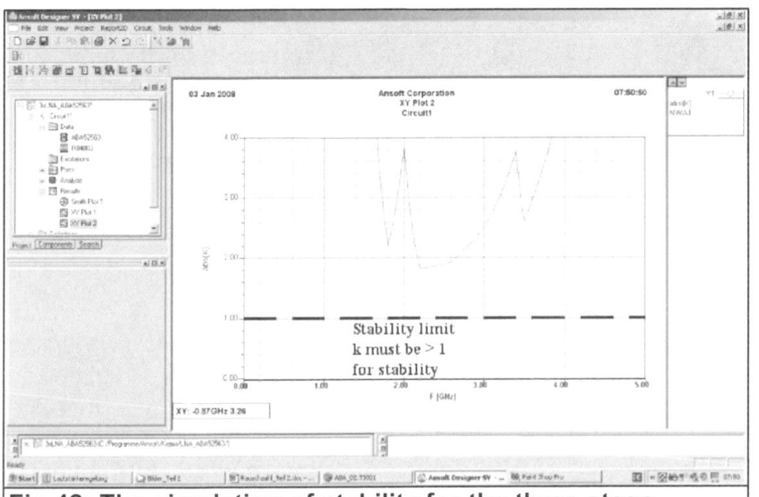

Fig 42: The simulation of stability for the three stage amplifier.

output and then the amplifier examined under different conditions. Here are the results:

- If the aluminium cover of the housing is omitted, the circuit does not oscillate, but there was a strong peak in the output noise at approximately 1.5GHz .
- Presenting the blank aluminium cover was sufficient to bring the circuit to wild oscillation.
- If absorbing foam material was stuck onto the inside of the cover the amplifier looked nearly perfect; the oscillation was completely gone and the frequency response of the noise on the screen was nearly even. Three different materials were tried but the jet-black material used for the packing ESD sensitive semiconductors gave very good results.
- Finally a professional microwave absorption material from Cummings was tried. This is designed for use up to 24GHz and was stuck on the inside of the cover (thanks to Dieter, DL6AGC, for providing a sample). The oscillation and the instability in the frequency response had almost disappeared. As a suspicious person, the 50Ω resistor was removed from the input. A 1 – 2db rise in the noise at 1.5GHz resulted. In order to cure that completely a 1dB SMD attenuator can be permanently fitted to the input, however the self-noise figure of the amplifier increases by 1dB.

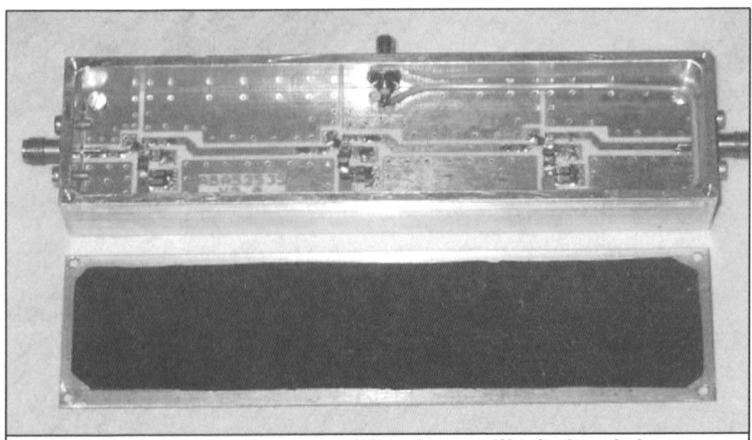

Fig 43: The three stage amplifier in a milled aluminium enclosure with absorption material on the lid.

In order verify our own measurements, a modern Agilent network analyser was used to measure the S parameters at a friendly company. Due to the high gain of over 60dB a 20dB SMD attenuator had to be attached to the input of the test specimen in order to avoid overloading the measuring input. Fig 44 shows the result, the curves for S21 and S22 supply the following information:

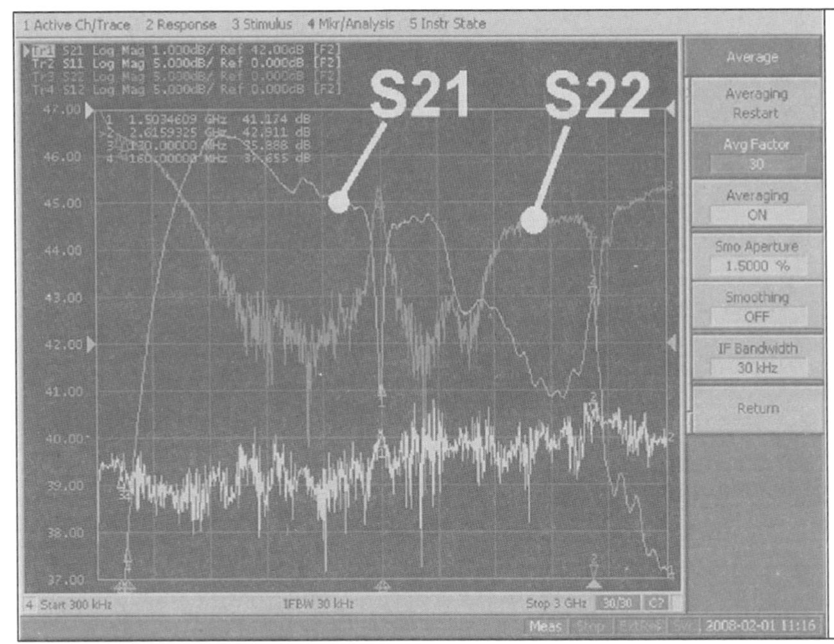

Fig 44: Measurements with an ultra-modern network analyser. See text for details.

- The gain is expressed by the parameter S21 with the scale from +37 to 47dB on the left axis (20dB must be added because of the attenuator) measures more than +60dB from approximately 200MHz to 2700MHz with a maximum value of +66.5dB at approximately 650MHz.
- At the sensitive oscillation point of 1500MHz a notch of approximately 3dB can be seen, this is nearly always an indication for a self-resonance probably the 33nH choke.
- S22 (right axis) shows a blip at this frequency. At lower frequencies the reassuring value of –20dB is measured, this only worsens above 2100MHz up to –15dB and finally at 3GHz a maximum of –10dB.

The result was not perfect for the first prototype, but nevertheless for investigations and measurement of noise factor it will be quite usefully.

To use the amplifier first we have to find the suitable settings for the spectrum analyser. For optimum at operation, accuracy and readability manual scan proved most suitable when the analyser works as tuneable receiver. Now ensure that the self-noise of the analyser remains at least 10 to 20dB below the output noise of the preamplifier by using as low an amplification of the prescaler in the analyser as possible. At the same time keep an eye on the overload limit of the analysers mixer input. Select the 10Hz video filter and a scale factor of 2dB per division to get a fine bright point screen display that varies irregularly by approximately 0.2dB. Now you can relax for the first time and collect the following measurements:

- Removing and replacing the 50Ω resistor on the input changes the indicated output noise level by approximately 2dB.

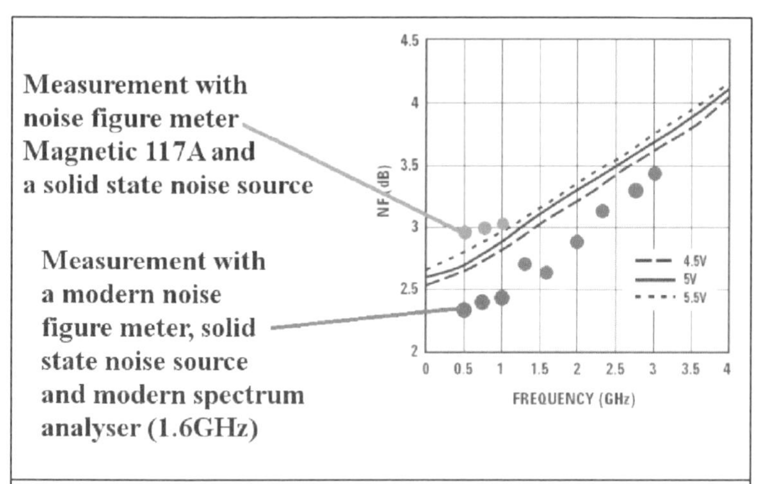

Fig 45: The hour of truth: Measurements with professional noise figure meters.

- With the 50Ω resistor fitted warming it by hand or violently with a soldering iron shows the forecast rise of the noise level very clear since the resistor now has a noise temperature of the soldering iron that is more than 290° Kelvin.

Thus the development phase is finished and genuine measurements can be made.

The single stage amplifier prototype as the Device Under Test (DUT)

The frequency range to 1.5GHz was selected because (owing to the help of the friend already mentioned) accurate measurements with modern professional instruments were available. Fig 45 shows these results of measurement compared with the data from the data sheet.

The three-stage prototype amplifier with more than 60dB gain was screwed onto the analyser, the 50Ω resistor fitted to the input, the centre line centred and only then the item under test with approximately 20dB gain was fitted. As expected the noise level rose by around approximately 20dB between 300MHz and 1.5GHz (according to the method described earlier). Instead of the expected NF of approximately 2.5 to 3dB (see Fig 45), a NF of less than 0.5dB was measured, although the S21 of DUT was known from the network analyser measurements for all selected measuring frequencies with an accuracy of approximately 0.2dB.

What had gone wrong? Interestingly the results of measurements from 2.5 to 3GHz converged slowly with the factory specifications according to Fig 456!

It is easy to be smart after the event, but the way to realisation was laborious despite the moral support of intensive discussions with competent people. Naturally the following rough calculations can be made:

The approximate middle noise output level in dBm of the chain of four prototype stages that feed the analyser input:

P_{noise} = (spectral noise density per Hz) + (increase by larger bandwidth) + (overall gain of all amplifiers) + (middle noise Figure of the first stage)

That results in the following total level in dBm:

P = -174dBm/Hz + 10 x log (2.5GHz) + 84dB + 3dB = -174dBm + 94dB + 84dB + 3dB = +7dBm

The higher values of gain (of over 22dB per individual stage) between 500 and 1000MHz were not considered at all, and if one then looks at the data sheet:

P1db = +9.8dBm

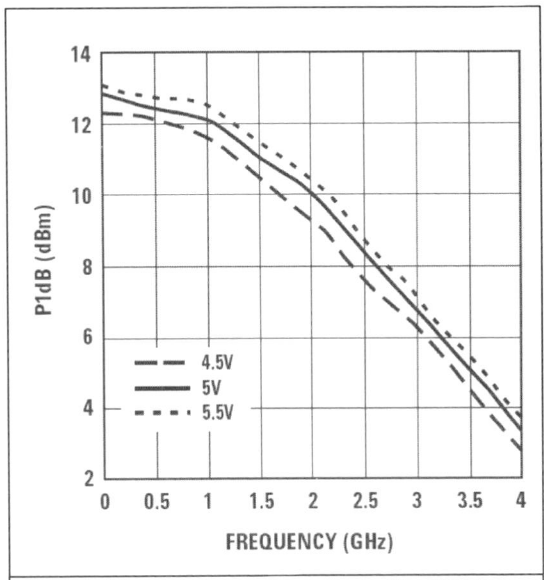

Fig 46: Here one can see how the MMIC output stage goes to its knees with rising frequency. Therefore the wrong results were obtained from the test run!

It becomes clear: P1dB defines the level at which at the output amplitude (by limiting) is already reduced by 1dB and this affects the last amplifier stage. The larger voltage peaks naturally occurring in the noise suffer and they are cut off. Fig 46 supports this realisation from the data sheet of the ABA52563, which shows the reduction of the P1dB value with rising frequency.

Since this method is not valid the following work is required: A further PCB must be designed but this time with only 2 stages and a sufficient overall gain between 40 and 45dB. The following improvements will follow:

- The microstrip lines on the PCB are replaced with "Coplanar Waveguides". These resemble microstrip lines, but ground surface is much closer to the left and right of the line. This leads to substantially smaller scattering fields and improves stability. Designing the lines is not a problem because the line calculator in the Ansoft designer student version has all conceivable models and options ready, free of charge.
- A copper sheet is inserted between the two stages, soldered to the ground surface as a partition to reduce the crosstalk between the stages. The through hole that connects the two stages must be very small due to the Coplanar lines.
- Different values for the 33nH choke on the output of each stage are chosen e.g. 27nH and 33nH. In this way there is no longer the bandpass filter amplifier effect with the periodic resonances. A bandpass filter amplifier relies on resonant circuits that are all at the same frequency, this is avoided with this detuning.
- The absorption material stuck on the inside of the lid cover must not be forgotten. The copper partition should exert some pressure on the foam material so that a gap is avoided.

The three-stage amplifier prototype has not been thrown away. With a low-noise, narrow-band preamplifier and additional band limitation with filters (to the avoidance of the limiting effect described above) you can hear the grass grow on a suitable Patch or Yagi antenna.

Summary

Once again much has been learned, not only about the phenomenon of the noise and noise measurement, prototype design of LNAs with a large frequency range and high gain and the fight with the self-oscillation. Also becoming acquainted with the spectrum analyser how to set the controls and switches in order to make the correct measurements. Naturally this gives food for thought.

In conclusion a small anecdote, in order to show that the phrase "nobody is perfect" will probably always apply:

After everything was clear with a good conclusion and the follow up procedure, it suddenly occurred to me that I had bought and stored four identical but defective Noise Figure meters at least 10 years ago from a HAM Radio Flea market. These were model 117A from the Magnetic company of Sweden constructed in 1970 including an associated solid state noise source for the frequency range to 1000MHz. After the question: "where did I put them" an intensive search followed and the answer was "in a heap with other electronics scrap iron in the garage" The next question was: "where is the service manual" once it was found this was followed by faultfinding. After appropriate expenditure of time now two of the four devices run again and third is waiting for two officially no longer available adjustable voltage regulators for the power supply that have been found using the Internet residing in the USA. The large meter in the fourth instrument is hopelessly defective and destroyed. So this set is only useful as a source of spare parts.

Excitingly after switching on and the comparing with the measurements made by the friendly company in short; the two antique devices measured their 1000MHz NF values of the small and large preamplifiers that were only around 0.3dB over the data sheet curves (see Fig 45). What more does one want? Which one to use now with so many devices?

Another short remark from the author

With this topic I have released things like an avalanche, because the results after the original publication of this article were some discussions by email and by telephone. Fortunately all were friendly and enriching for my own knowledge, uncovering my own mistakes in reasoning or simply becoming acquainted with very nice and competent colleagues.

Therefore I thank everyone for the lively reaction. Thanks also for all the additional information about things like the fast aging noise diodes that are probably still to be found in America on the Internet as a spare part at reasonable prices. And naturally for all practical assistance with measurements and discussions during the project.

References

[1] Improving harmonic frequency measurements with the HP8555A, Ralph Berres, DF6WU, VHF Communications Magazine 1/2009 pp 25 - 36

[2] 10MHz - 10GHz diode noise source, Franco Rota, I2FHW, VHF Communications Magazine 4/2008 pp 241 - 248

[3] VHF Communications 1/2007 "Noise source diodes"

[4] For those who need more information about the mismatch uncertainty in noise figure measurement I suggest 3 application notes:

- Ham Radio, August 1978
- Noise figure measurement accuracy AN57-2 Agilent
- Calculating mismatch uncertainty, Microwave Journal May 2008

[5] R.F. Elettronica web site catalogue www.rfmicrowave.it (capacitors section)

[6] VHF Communications 4/2004 "Franco's finest microwave absorber"

[7] The RF Elettronica catalogue - www.rfmicrowave.it/pdf/diodi.pdf (from page A 14)

[8] Practical project: Noise factor measurements with older spectrum analysers, Gunthard Kraus, DG8GB, VHF Communications Magazine 1/2008 pp 2 - 11 and 3/2008 pp 140 - 154

[9] David Pozar, "Microwave engineering", Prentice Hall, Englewood Cliffs, N.J. 07632. ISBN 0-13-586702-9

[10] HP Agilent Application note 57, parts 1/2/3 (Download from the Web)

[11] HP Agilent Application note 150, parts 1/2/3/4 (Download from the Web)

[12] HP Agilent Application note 1354: "Practical Noise Figure Measurement and analysis for Low Noise Amplifier Design" (Download from the Web)

[13] Maxim Application note 2875: Three Methods of Noise Figure Measurement (Download from the Web)

[14] Mohr Associates, Inc.: Tutorial "Mohr on Receiver Noise". (Download from the Web)

[15] PUFF CAD software - available from www.vhfcomm.co.uk

[16] Ansoft Designer SV (Student Version) - http://www.ansoft.com/ansoftdesignersv/. Ansoft Corporate Headquarters, 225 West Station Square Drive, Suite 200, Pittsburgh, PA 15219, USA, Tel: (412) 261-3200, Fax: (412) 471-9427

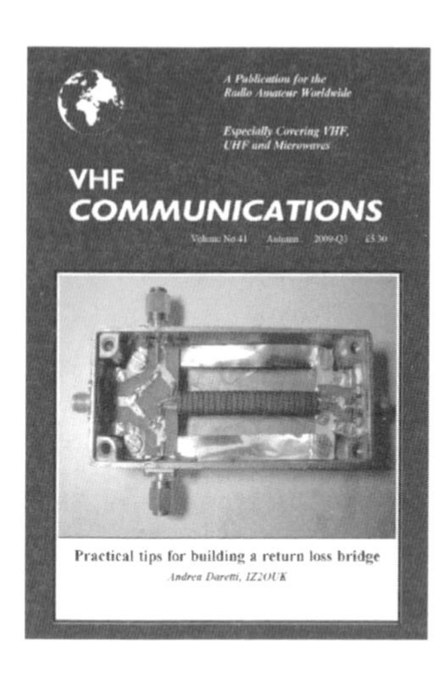

VHF Communications Magazine

- Constructional and theoretical articles on all aspects of VHF, UHF and Microwaves
- Published in English since 1969
- Available by annual subscription
- Kits and PCBs available for most constructional articles

Web site www.vhfcomm.co.uk

- Subscribe on line
- List of overseas agents
- Downloads of some past articles
- Search or download a full index of the magazine since 1969
- Links to other interesting sites

Back Issues

- All back issue available as real magazines or photocipies
- Back issues on DVD, a whole decade of magazines on each DVD
- Binders to hold 12 issues (3 years of magazine)

K M Publications

63 Ringwood Road, Luton, Beds, LU2 7GB, UK

Tel/Fax: +44 (0) 1582 581051 email: andy@vhfcomm.co.uk

Chapter 4

Filters and Design

In this chapter :

- Using microstrip interdigital capacitors
- Bandpass filter for microwave projects
- A Low Loss 13cm High Power 90° Hybrid Combiner

Filters are very important to ensure that your transmitters have a clean output and that your receivers are listening to what you think they are. Designing filters for the microwave bands presents its own problems but with modern techniques these problems can be solved by most radio amateurs. This chapter shows some of these techniques and some very useful designs.

Using microstrip line interdigital capacitors [1]

Very small coupling capacitors are required for bandpass filters in the frequency range between 100MHz and 1GHz, often with values under 0.5pF. Implementing these as microstrip interdigital capacitors in microstrip gives some advantages. This will be demonstrated in the following practical development.

As an introduction into microstrip interdigital capacitors (Fig 1), an extract from the on-line help of the CAD program gives all the necessary explanation and details.

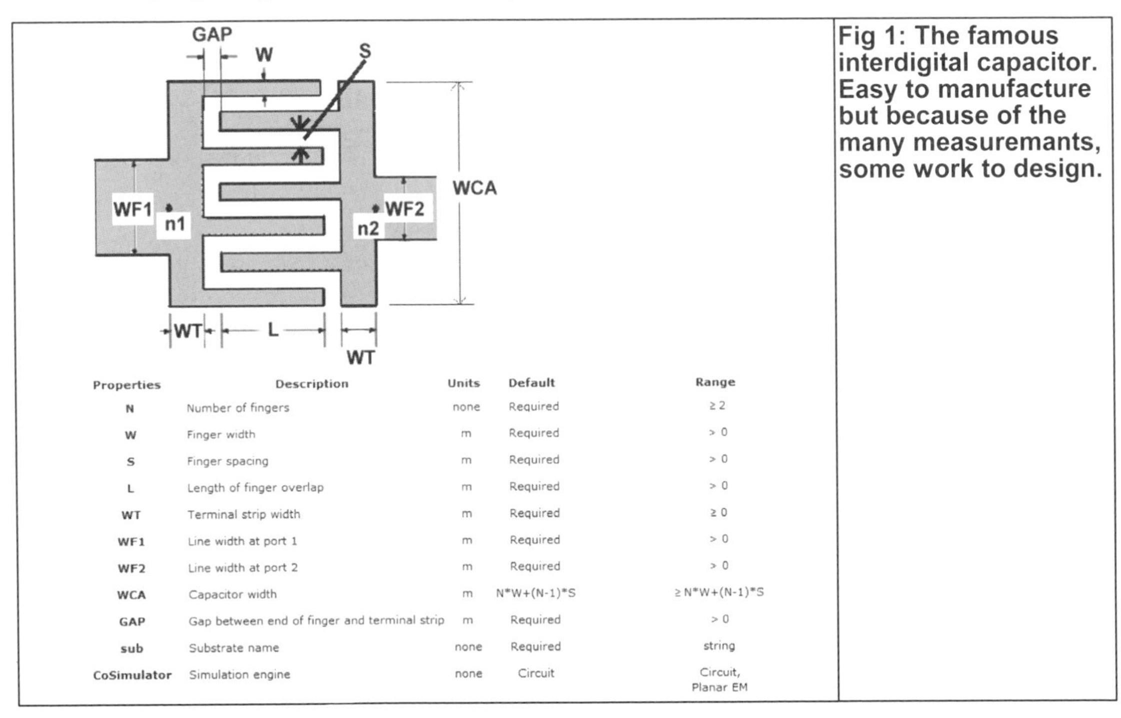

Properties	Description	Units	Default	Range
N	Number of fingers	none	Required	≥ 2
W	Finger width	m	Required	> 0
S	Finger spacing	m	Required	> 0
L	Length of finger overlap	m	Required	> 0
WT	Terminal strip width	m	Required	≥ 0
WF1	Line width at port 1	m	Required	> 0
WF2	Line width at port 2	m	Required	> 0
WCA	Capacitor width	m	N*W+(N-1)*S	≥ N*W+(N-1)*S
GAP	Gap between end of finger and terminal strip	m	Required	> 0
sub	Substrate name	none	Required	string
CoSimulator	Simulation engine	none	Circuit	Circuit, Planar EM

Fig 1: The famous interdigital capacitor. Easy to manufacture but because of the many measuremants, some work to design.

The design is not simple, however modern microwave CAD programs facilitate simulation; these should already contain this component in their component library as a microstrip line model.

That is the case for the free Ansoft Designer SV software, this is the list of the advantages:

- After optimisation of the PCB layout very small tolerances are achieved leading to good reproducibility of filter parameters without additional components or assembly costs for quantity production.
- No discrete components need to be soldered. These would be difficult to obtain for such small capacitances and exhibit larger tolerances.
- Using high quality printed circuit board material with the smallest losses produces very high quality capacitors that are useful up to more than 10GHz.

The project, a 145MHz bandpass filter

A bandpass filter with the following data is to be designed, built and measured:

- Centre frequency: 145MHz
- Ripple bandwidth: 2MHz
- System resistance Z: 50Ω
- Filter degree: n = 2
- PCB size: 30mm x 50mm
- Tschebyschev narrow bandpass filter type with a Ripple of 0.3dB (coupled resonators)
- PCB material: Rogers RO4003, thickness: 32MIL = 0.813mm, $\varepsilon_r = 3.38$, TAND = 0.001
- Housing: Milled aluminium
- Connection: SMA plug

First design:

- Filter coils NEOSID (type 7.1 E with shielding can, L = 67 - 76nH, single coil, quality Q = 100 -150, brass adjustment core)
- SMD ceramic capacitors 0805, NP0 material

The filter program contained in Ansoft Designer SV was used. The development of the circuit after the draft and a short optimisation is shown in Fig 2. The further work necessary to produce the finished PCB layout is described in the following article. A prototype was produced and tested using a network analyser to give the measurement results.

The design of the filter using Ansoft Designer SV giving all the steps leading to Fig 2 are shown

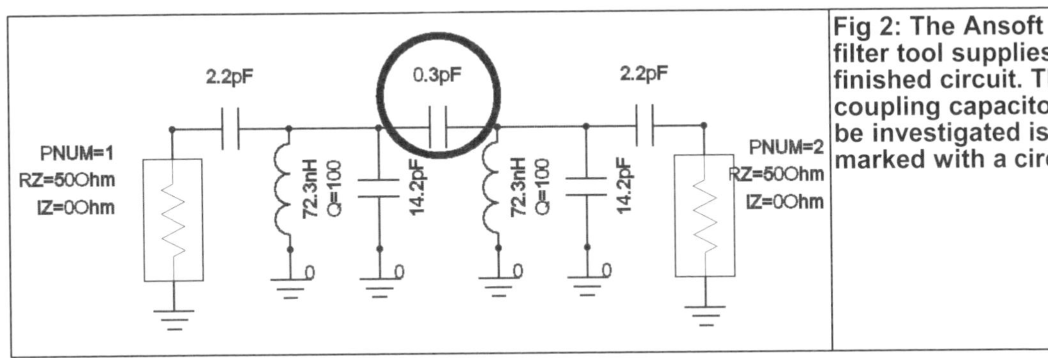

Fig 2: The Ansoft filter tool supplies the finished circuit. The coupling capacitor to be investigated is marked with a circle.

in Appendix 1.

Appendix 2 contains guidance for successful control of the circuit simulation using Ansoft Designer SV.

It will also be helpful to download a copy of the authors tutorial on using Ansoft Designer SV. This is available free of charge in German or English from the web site [2].

To continue with the filter development; a look at the circuit of Fig 2 shows:

- The problematic coupling capacitor C = 0.3pf is identified by the black circle. The problem is not only the very small capacitance but also the high requirement for accuracy. A deviation of more than 1% gives a noticeable change in the transmission characteristics.
- It was optimised until all remaining capacitors can be realised using standard values, if necessary by parallel connection of several capacitors with different values.

Design procedure for interdigital capacitors with Ansoft Designer SV

Input problems

The component in the model library is under:

- Circuit Elements/Microstrip/Capacitor/MSICAPSE

The layout of the interdigital capacitor in series connection is shown in Fig 1. Double click on the circuit symbol to access the list of the dimensions. At the end of the list there is a "MSICAP" button that opens the on-line help with an explanation of the individual inputs and dimensions. Experience is required to make these inputs but if the following rules are used then incorrect inputs will be avoided:

- Set the finger width W to 0.5mm. This ensures that design does not become too large and under etching has less effect on the finger width when the PCB is made.
- The gap width S should not be TOO small otherwise the PCB manufacturer complains. A value of 0.25mm can be achieved even in your own workshop. On

the other hand it should not be too large because then the capacitance value falls requiring larger finger lengths or more fingers.

- The number of the fingers and their length specifies the capacitance value. As an example, start with 4 fingers and vary the length of the fingers ensuring that they do not become too large. Set an upper limit of about 8 to 10mm. Instead of making the fingers longer simply increase the number of fingers.

With this data (and the PCB data) a draft design can begin BUT unfortunately the student version of the CAD program can only make one analysis.

That means that all the data and dimensions can be entered and the simulation started but the result will be an S parameter file of the capacitor. It is only at this point that it is known if the capacitance of the draft capacitor is too large or too small.

Things become more difficult because the circuit diagram of the component has additional capacitors from each end to earth. These two unavoidable parallel capacitors detune the resonant circuits. How can these three capacitances be isolated to optimise the circuit, particularly if the filters are more complex and several interdigital capacitors are used?

Determination of the pure capacitance

It is a challenge to determine the exact value of the coupling capacitor, but the two parallel capacitors are less difficult to deal with, just make the resonant circuit capacitors smaller in the simulation until the desired transmission curve is achieved. The difference corresponds to the additional parallel capacitance contributed by the interdigital capacitor. Their value is not much different from the actual coupling capacitor.

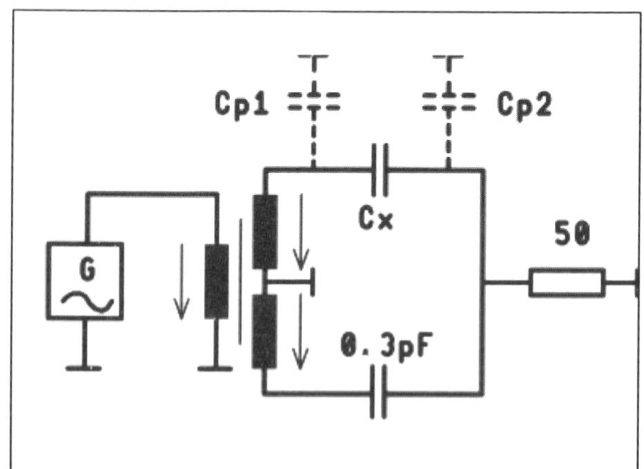

Fig 3: Using an idea from crystal filter technology, this circuit is used to develop the exact value of the interdigital capacitor required.

Sometimes experience helps: e.g. crystal filters in a bridge connection presented a similar problem. In that case the housing capacitance was eliminated using a transformer circuit. The principle applied to the current problem is shown in Fig 3.

The voltage across the two secondary windings are the same magnitude but opposite phases. Thus the voltage, V, across the terminating resistor, RL, is zero if the two capacitors Cx and C2 are equal and in this case equal 0.3pF. The two parallel capacitances Cp1 and Cp2 do not play a role when the bridge is balanced thus only the value of Cx is being measured. Cp1 is parallel to the secondary winding of the transformer and cannot affect the balance of the bridge. Likewise Cp2 is in parallel with the 50Ω load resistor. In the balanced condition no voltage is developed across the parallel capacitors and therefore Cp2 has no effect on the circuit. This means that the value of interdigital capacitor can be simulated to be exactly the same as the known capacitor C2 and its

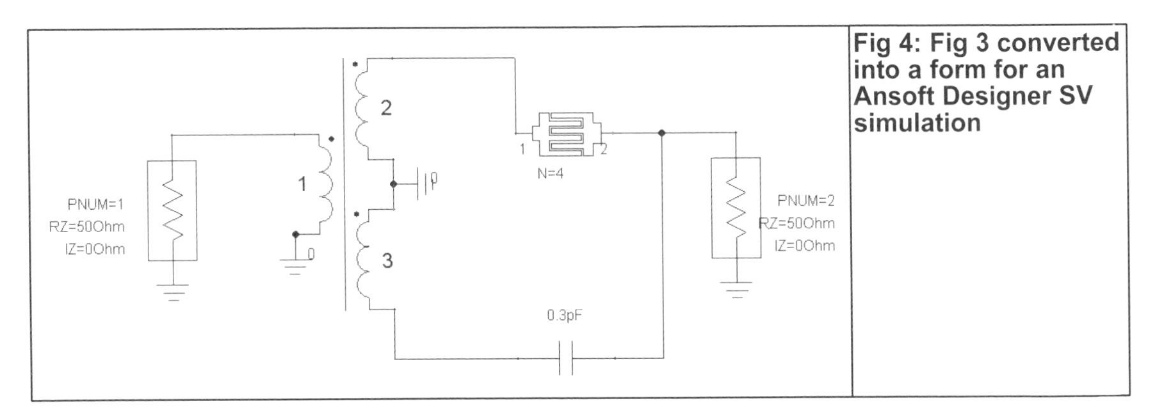

Fig 4: Fig 3 converted into a form for an Ansoft Designer SV simulation

parameters will then be known independent of the parallel capacitance values.

The simulation circuit shown in Fig 4 can be developed using Ansoft Designer SV. The transformer can be found in the component library under:

- Components/Circuit Elements/Lumped/Transformers/TRF1x2

and the series connected interdigital capacitor under:

- Components/Circuit Elements/Microstrip/Capacitors/MSICAPSE

A microwave port with internal resistance of 50Ω feeds a broadband transformer with two secondary windings. The upper coil is connected to the output port by the interdigital capacitor. The opposite phase signal supplied by the lower coil is fed via a second capacitor to the output port.

Importantly:

This second capacitor must have the same value of 0.3pF (value of the interdigital coupling capacitor).

The important data for the simulation (and the later draft layout) is entered in the Property Menu of the interdigital capacitor. Double clicking on the symbol in the circuit diagram can opens the menu; the entries required are shown in Fig 5.

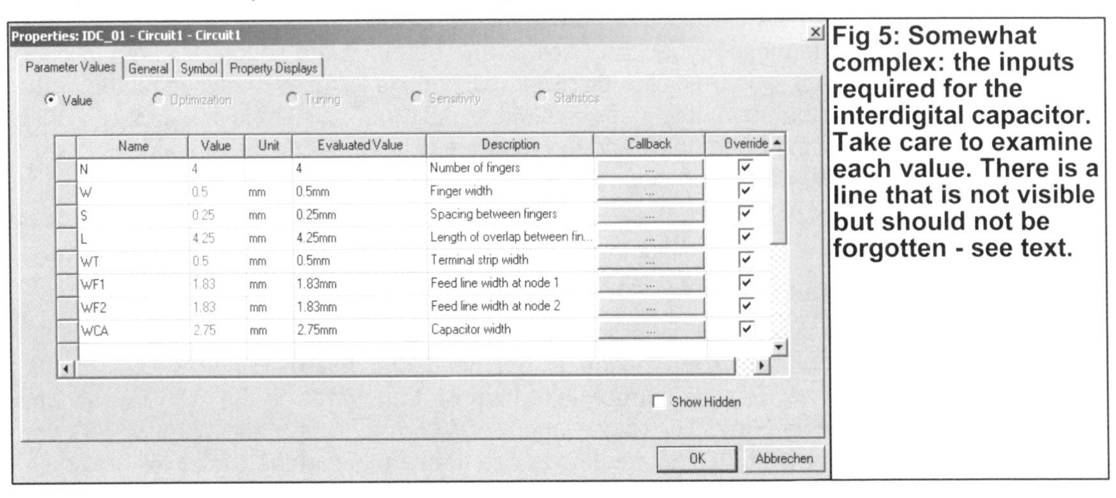

Properties: IDC_01 - Circuit1 - Circuit1

Parameter Values | General | Symbol | Property Displays

Value | Optimization | Tuning | Sensitivity | Statistics

Name	Value	Unit	Evaluated Value	Description	Callback	Override
N	4		4	Number of fingers	...	✓
W	0.5	mm	0.5mm	Finger width	...	✓
S	0.25	mm	0.25mm	Spacing between fingers	...	✓
L	4.25	mm	4.25mm	Length of overlap between fin...	...	✓
WT	0.5	mm	0.5mm	Terminal strip width	...	✓
WF1	1.83	mm	1.83mm	Feed line width at node 1	...	✓
WF2	1.83	mm	1.83mm	Feed line width at node 2	...	✓
WCA	2.75	mm	2.75mm	Capacitor width	...	✓

Show Hidden

OK | Abbrechen

Fig 5: Somewhat complex: the inputs required for the interdigital capacitor. Take care to examine each value. There is a line that is not visible but should not be forgotten - see text.

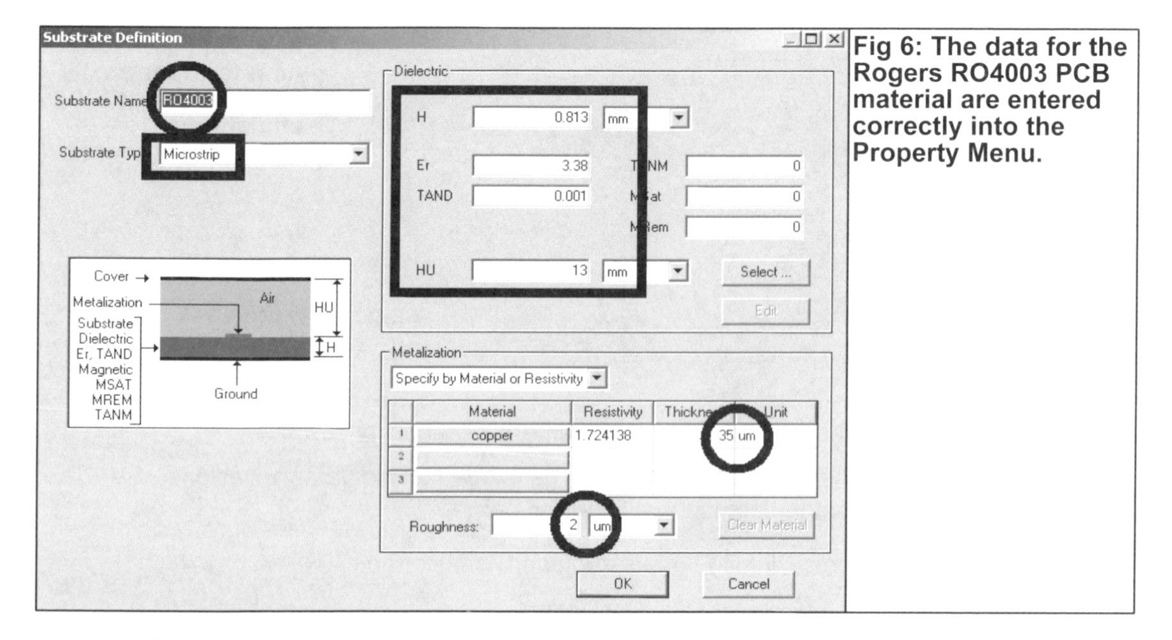

Fig 6: The data for the Rogers RO4003 PCB material are entered correctly into the Property Menu.

However there are two important things that are not immediately obvious:

- There is a line missing from the window shown in Fig 5, this can be found by scrolling down. The line to find is: GAP (between end of finger and terminal strip) = 0.25mm
- The total width "MCA" must be calculated by hand and entered into the relevant field. It should be noted that the units do not automatically default to mm so take care to enter this value otherwise the default will be metres and the simulation will be meaningless.

The above dimension is calculated as follows: MCA = 4 x Finger length + 3 x Gap width = 4 x 0.5mm + 3 x 0.25mm – 2.75mm

Finally the correct PCB material data must be selected. Scroll to the line "SUB" in the open Property Menu for the interdigital capacitor. Click on the button in the second column to open the menu "Select Substrate" and then click on "Edit". Fill out this form, as shown in Fig 6, for the RO4003 PCB material to be used: thickness = 32MIL = 0.813mm, dielectric constant ε_r = 3.38, tand = 0.001, copper coating 35μm thick and the roughness is 2μm. Once everything is correct click OK twice to accept the data and close the Property Menu.

Everything is now ready for the simulation. Program for a sweep from 100MHz to 200MHz with 5MHz increments and look at the results for S21 (If you do not know the individual input steps necessary for Ansoft Designer SV they are described in appendix 2).

The finger lengths of the interdigital capacitor are varied and the simulation run again until the minimum for S21 is found. Now the bridge is balanced and the mechanical data for the interdigital capacitor with exactly the correct value can be transferred to the PCB layout. The optimised results are shown Fig 7. The finger length of 4.25mm gives a minimum for S21 and further refinement is not needed. It is interesting to see the results that the circuit will provide.

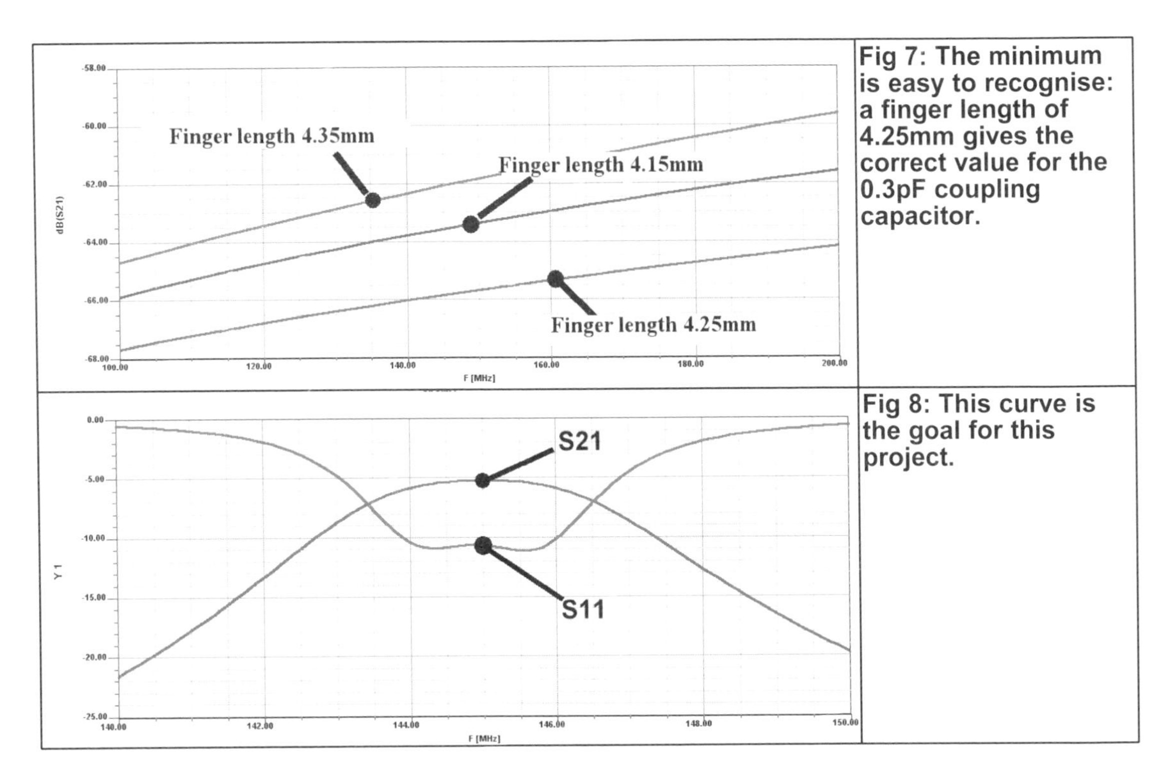

Fig 7: The minimum is easy to recognise: a finger length of 4.25mm gives the correct value for the 0.3pF coupling capacitor.

Fig 8: This curve is the goal for this project.

Completing the circuit

A new project is started with the circuit as shown in Fig 2 (with discrete components) and the sweep adjusted from 140 to 150MHz in steps of 100kHz with S11 and S21 displayed. This result is shown in Fig 8. This is the starting point for the following actions, if these result in the same result the you can be quite content.

Replacing the 0.3pF coupling capacitor with the interdigital component that has been designed gives the simulation circuit shown in Fig 9. Naturally the results shown in Fig 10 are worse because the additional parallel capacitances of the interdigital capacitor have not been considered. The parallel capacitors in both resonant circuits must be reduced until the curves of Fig 8 are achieved. Fig 10 shows an additional surprise that apart from the expected shift of the centre frequency from 145MHz to 143.3MHz (caused by the parallel capacitance of the

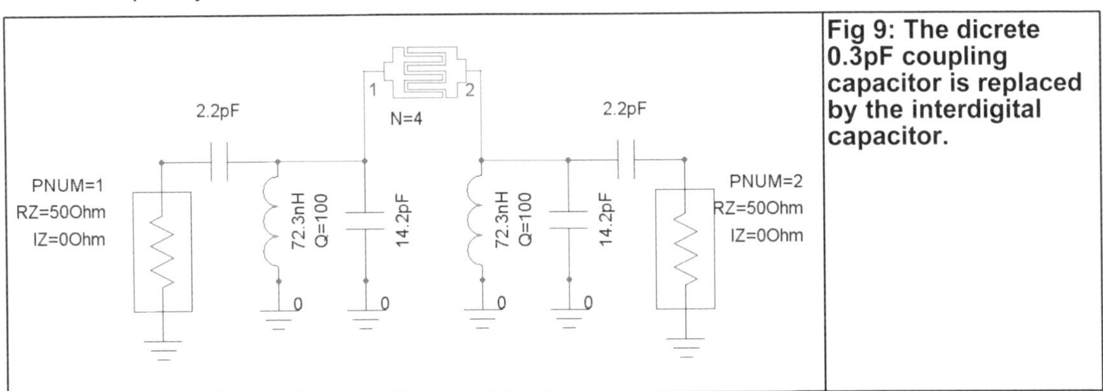

Fig 9: The dicrete 0.3pF coupling capacitor is replaced by the interdigital capacitor.

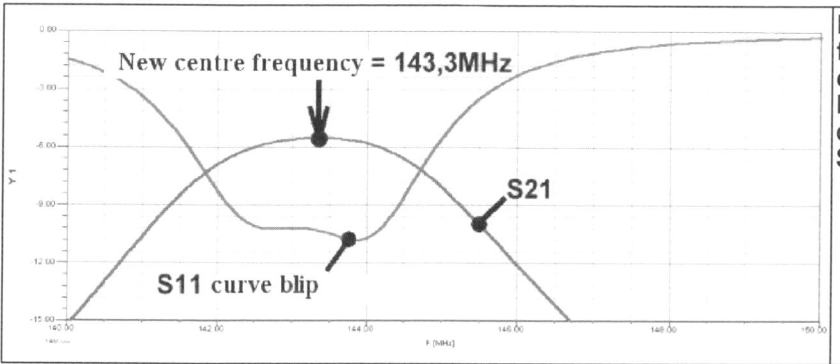

Fig 10: The centre frequency has, as expected, moved lower and there is a diagonal dip in the S11 curve.

interdigital capacitor) the S11 curve has a diagonal dip. Trying to compensate this effect with different values of two parallel inductances is surprising because as S11 improves, S22 gets worse. This means that this interdigital solution has its peculiarities based on the frequency response of the capacitors. Probably an alternative circuit diagram with only 3 capacitors can be imagined but the effect is more complex because it can be seen on the finished PCB. The effect can be lived with so the easy solution is just to move the centre frequency to the required value of 145MHz.

The parallel capacitors must be reduce to 138pf = 12pf + 1.8pf, using standard values that can be connected in parallel. The remaining adjustment is to fine tune the two coils using the adjustable cores. The new inductances of L = 72.3nH corresponds to the simulation result shown in Fig 10.

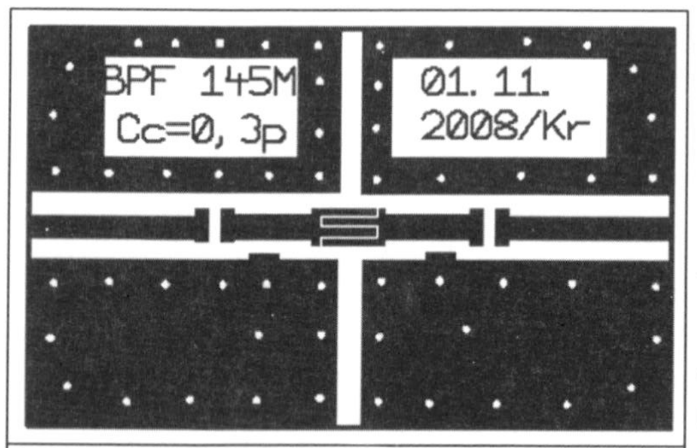

Fig 11: The printed circuit board measures 30mm x 50mm made from Rogers RO4003 with a thickness of 0.813mm.

But this is not the conclusion because the PCB layout and its influence must be considered. Fig 11 shows the PCB layout that is principally a 50Ω microstrip line. It starts on the left (at the input SMA connector) with a gap for the 2.2pF SMD coupling capacitor followed by the resonant circuit. The interdigital capacitor is in the centre with the right half being a mirror image of the left hand side. This corresponds to an additional conductor length of approximately 40mm for the circuit and this has the following consequences:

- Four additional sections of 50Ω microstrip line (with a width of 1.83mm for the given PCB data) must be added to the Ansoft Designer SV circuit if the simulation is to agree with the reality.
- The lengths of the pieces of line are 2 x 13mm = 26mm (from the SMA connector to the 2.2pF coupling capacitor) and 2 x 7mm = 14mm (from the resonant circuit to the interdigital capacitor).

Fig 12 shows the circuit. At these relatively low frequencies the microstrip line detunes capacitors so the parallel components must be adjusted again. Doing this gives the simulation

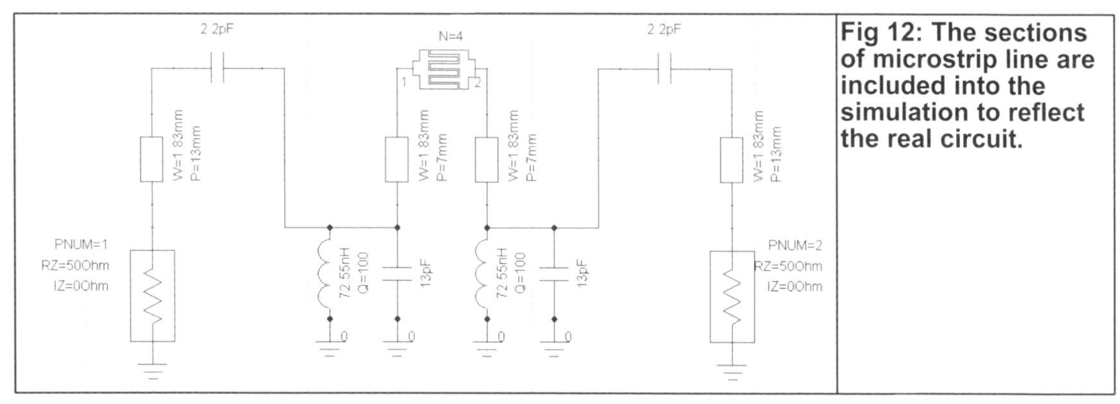

Fig 12: The sections of microstrip line are included into the simulation to reflect the real circuit.

results shown in Fig 13. The simulation of the wider frequency range from 100MHz to 200MHz is shown in Fig 14. Finally it is time to prepare the prototype PCB, the result after some hours of work are shown in Fig 15.

About 50 0.8mm hollow rivets were used for the plated through holes from the ground islands to the continuous lower ground surface. The SMD capacitors and coils are soldered and copper angles are screwed on to fit the SMA sockets. The adjustment cores of the coils are now easily accessible and from experience it is known that nearly no further adjustments are necessary when fitting the PCB into a machined aluminium housing.

The truth comes with the comparison of the curves of Fig 13 and 14 with the image that the network analyser produces from the prototype.

By the way: the tear on the PCB that can be seen in Fig 15 was caused by human error. It is

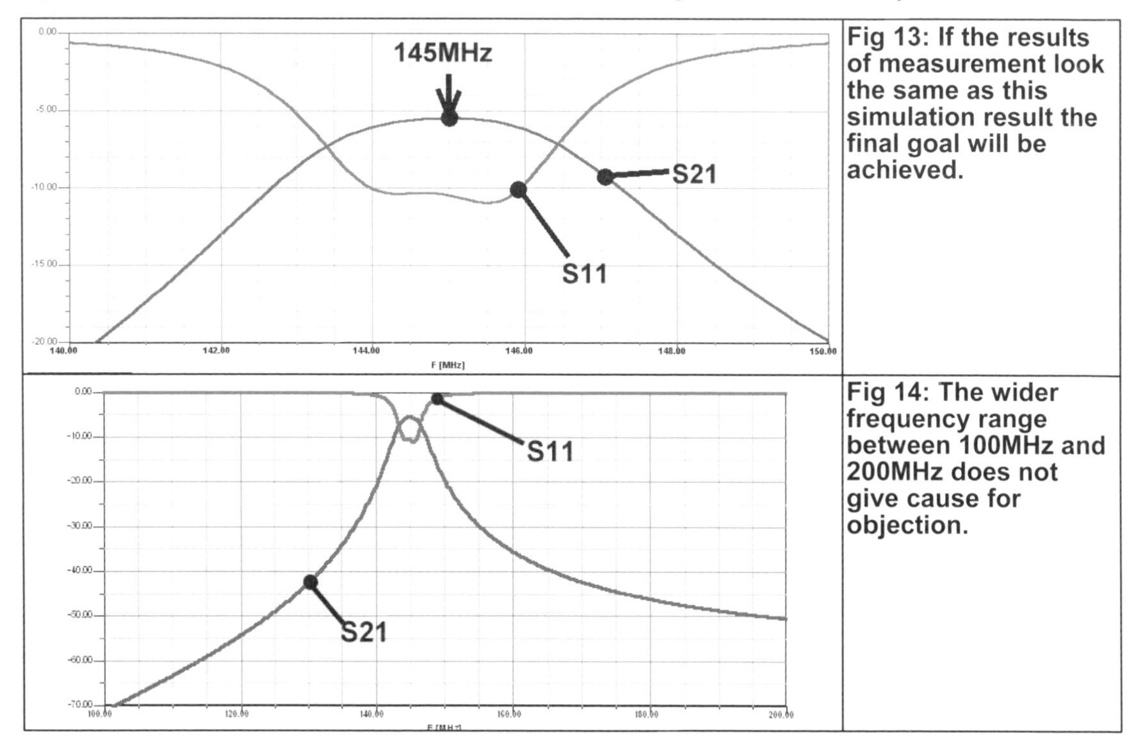

Fig 13: If the results of measurement look the same as this simulation result the final goal will be achieved.

Fig 14: The wider frequency range between 100MHz and 200MHz does not give cause for objection.

Fig 15: The simulations converted into a prototype with SMA connectors. The circuit must now be measured on the network analyser.

hard work fitting so many small rivets and takes some hours. But afterwards when finishing the PCB with a file in a hurry to see the results, too much pressure was applied. So you find out that RO4003 material can be drilled and milled but protests when it meets a stronger opponent. More care needed in future.

Results of measurement on the prototype

The measurements gave some unpleasant surprises shown in Fig 16, which is the S21 transmission curve (measured after correct alignment) and the simulation from Fig 13. The first mystery is that the attenuation has risen from 5.5 to 7dB at the centre frequency.

There are some doubts about the method devised to measure the value of the coupling

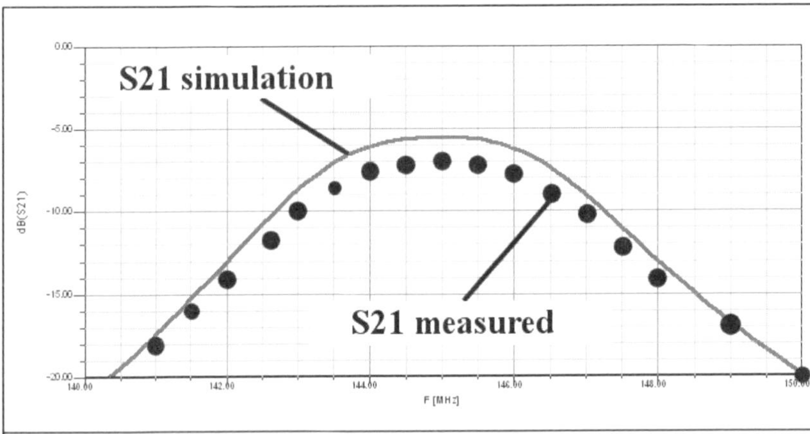

Fig 16: The measured S21 response shown with the simulation.

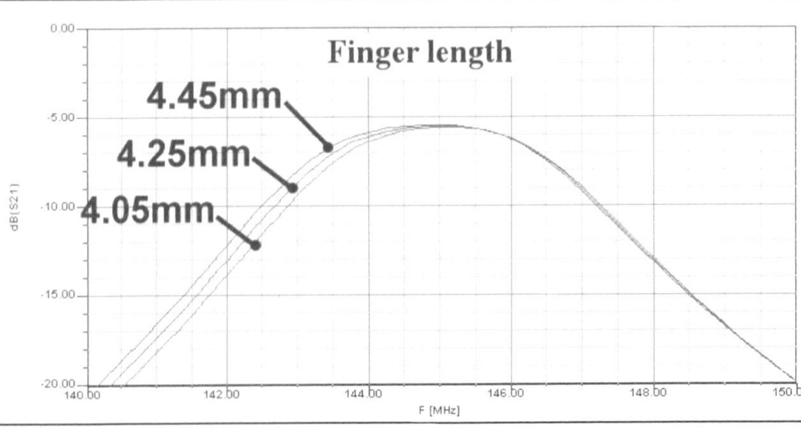

Fig 17: Different finger lengths and therefore different coupling capacitor values only move the centre frequency of the transmission curve but have no influence on the attenuation.

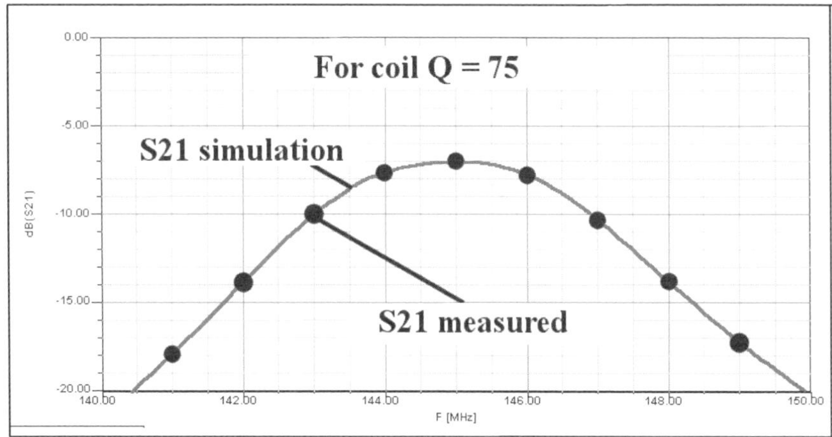

Fig 18: This proves the coils are the problem, the picture speaks for itself - see text.

capacitor even though the author is proud of the technique devised. so the best was to look for an owner of the APLAC simulation software (full version). APLAC has a text based command line simulator that can directly compute the value of the interdigital capacitor and the two "end capacitors". Entering the mechanical data for our capacitor design into APLAC and waiting for the result gave great relief because it gave a value of 0.29pF that is very close to the 0.3pF aimed at with Ansoft Designer SV. The interdigital capacitor is probably not the cause of the discrepancy but the sceptical developer leaves nothing to chance. The effect of changing the finger length by 0.2mm, and thus the coupling capacity, on the filter curve is shown in Fig 17. This gives the all-clear signal because the range of the transmission curve only changes slightly but the attenuation is not affected.

This leaves the parallel coils as the possible problem (once again) because the NP0 material used in the SMD capacitors is above suspicion at these frequencies. Therefore the coil quality must be worse than shown on the data sheet (Q = 130) and the reason could be because the inductance is adjustable using a brass core. Eddy currents induced in the core oppose the magnetic field to reduce the inductance but unfortunately the quality falls. The quality Q = 130 specified, only applies when the core is fully unscrewed and thus almost ineffective, giving the maximum inductance value. There is no mention of this in the data sheet.

This explains everything but to double check a further simulation was performed. Fig 18 shows the proof because with Q = 75 the simulation follows the measured S21 curve accurately. The measurements also agreed with the wide frequency sweep shown in Fig 14.

Summary

Interdigital capacitors are a fascinating component; as long as the PCB process has an accuracy of 0.01mm they are a good component for problem free mass production. Only the coils were a problem, more tests would be required to find a better solution.

After the prototype was built and discrepancies noticed the simulations served as an analysis tool to determine the cause of the errors. This was all at no cost and was fun to do. The author wishes that this has inspired you to use Ansoft Designer SV for your own projects, the appendices give more information for filter design.

Appendix 1: Help for using the filter program in the Ansoft Designer SV

There is no need to continue searching the Internet for suitable CAD software for filter design,

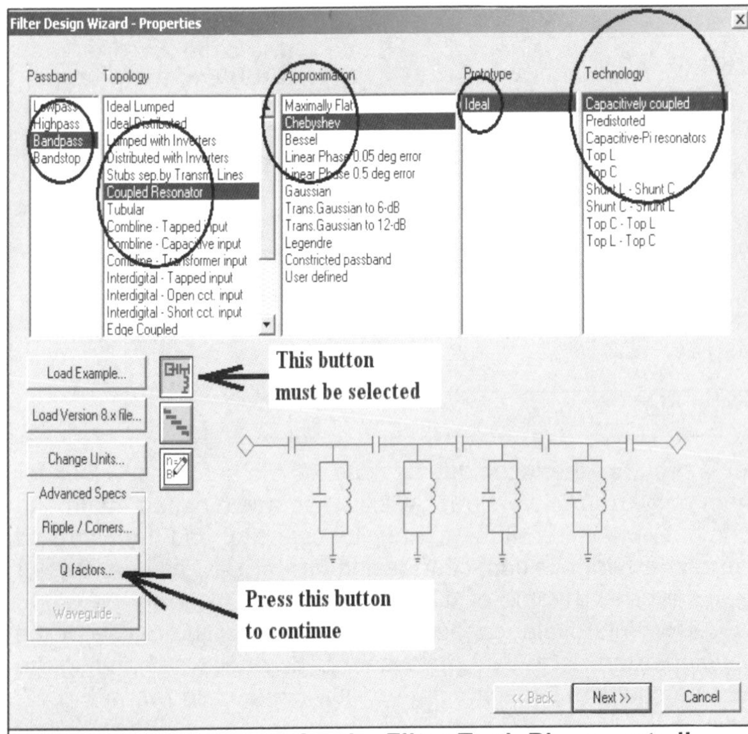

Fig 19: The start menu for the Filter Tool. Please set all values as shown.

because Ansoft Designer SV deals with almost every filter type possible. The problem is how to find the correct selection:

To start the designer with a new file and go to the "Project Open" option on the menu and click on "Insert filter Design". Then use something like:

Step 1 (see Fig 19):

In the five menus (from left to right) select:

- Bandpass/Coupled Resonator/Chebyshev/ Ideal/Capacitively Coupled

Select the button for "lumped design" (button with circuit diagram). If everything is done click on "Q factors".

Fig 20: The coils need special attention (as always). The quality is set as shown.

Step 2 (see Fig 20):

The coil quality is set to Qmin = 100 at 100MHz (the filter quality rises linear with frequency). Click OK to return to the previous screen and then click "Next".

Step 3 (see Fig 21):

Now for the serious entry of the filter data:

Order (= filter degree):	2
Ripple:	0.3dB
fp1 (lower cut off frequency):	0.144GHz
fp2 (upper cut off frequency):	0.146GHz
fo (centre frequency):	0.145GHz
BW (Bandwidth)	0.002GHz
Source, Rs (source resistance):	50Ω
Load, Ro (load resistance)	50Ω
Inductor L:	73nH

(selected parallel inductance, all the same)

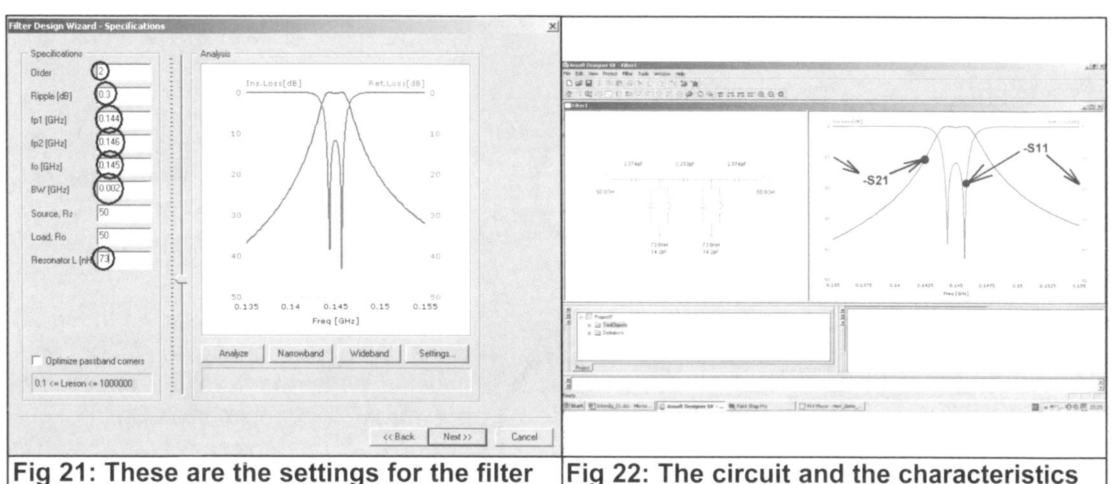

Fig 21: These are the settings for the filter and should be copied exactly - see text.

Fig 22: The circuit and the characteristics of the ideal filter.

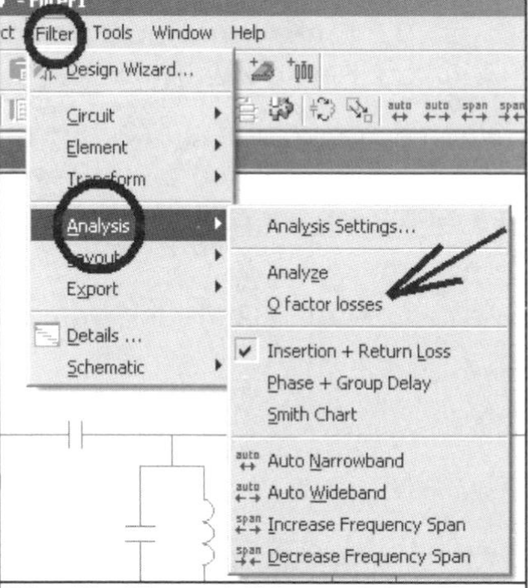

Fig 23: The addition of the filter quality will distort the characteristics.

Press "Next" and the circuit is produced, then select "Finnish".

Step 4:

Click "Tile vertically" to produce a display of the circuit diagram and associated simulation of S11 and S22 as shown in Fig 22. The vertical axis is marked with "Insertion Loss (dB)" and "Return Loss (dB)". S11 and S22 are obtained by reversing the sign of these.

Step 5:

To show the effect of the coil quality Q = 100 select the view shown in Fig 23 by using the "Filter" menu from the Filter 1 window border and select "Analysis" then "Q Factor Losses". If a check mark is set then Fig 24 shows the filter characteristics adjusted for the quality Q = 100.

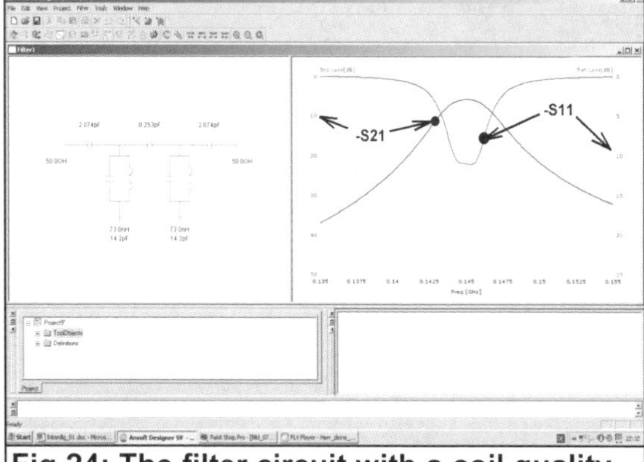

Fig 24: The filter circuit with a coil quality Q = 100.

Print the circuit and place it beside the PC because the next appendix needs the component values.

Appendix 2: Simulation of the circuit with the Ansoft Designer SV

Start a new project using the "Insert Circuit Design" option. The "Layout Technology Window" shows: MS-FR4 (Er=4.4), 0.060 inch, 0.5oz.copper,

At first, place the two ports required. Initially they are interconnected ports, double clicking on their circuit symbols give the chance to

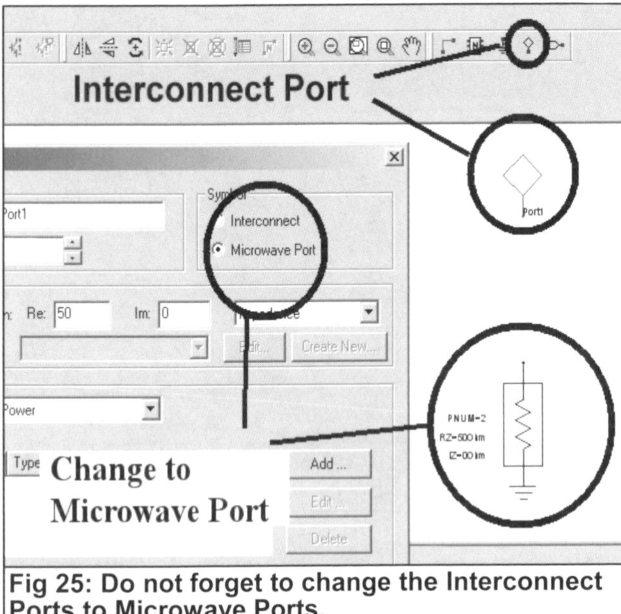

Fig 25: Do not forget to change the Interconnect Ports to Microwave Ports.

change them to Microwave Ports (Fig 25).

Now the remaining components can be found in the Project Window under the "Components/Lumped". For the capacitors a simple "Capacitor" is used but for the coils "INDQ" (Inductor with Q factor) should be used.

The circuit is drawn as shown in Fig 26 using "Wire" to connect the components and the component values added. Do not forget to double click on the coil symbols and set the quality to Q = 100 at 0.1GHz.

The PCB material should be changed to "32MIL = 0.813mm thickness and R04003 material" as shown in Fig 6.

Note: When a component is attached to the cursor it can be rotated by pressing "R". If the component is already placed it can be selected with a single mouse click and rotated by pressing "Control" and "R".

The circuit is stored under a suitable name and a sweep for 140 - 150MHz in 100kHz steps carried out as shown in Fig 27:

Step 1:

Click to the setup button and continue with "Next".

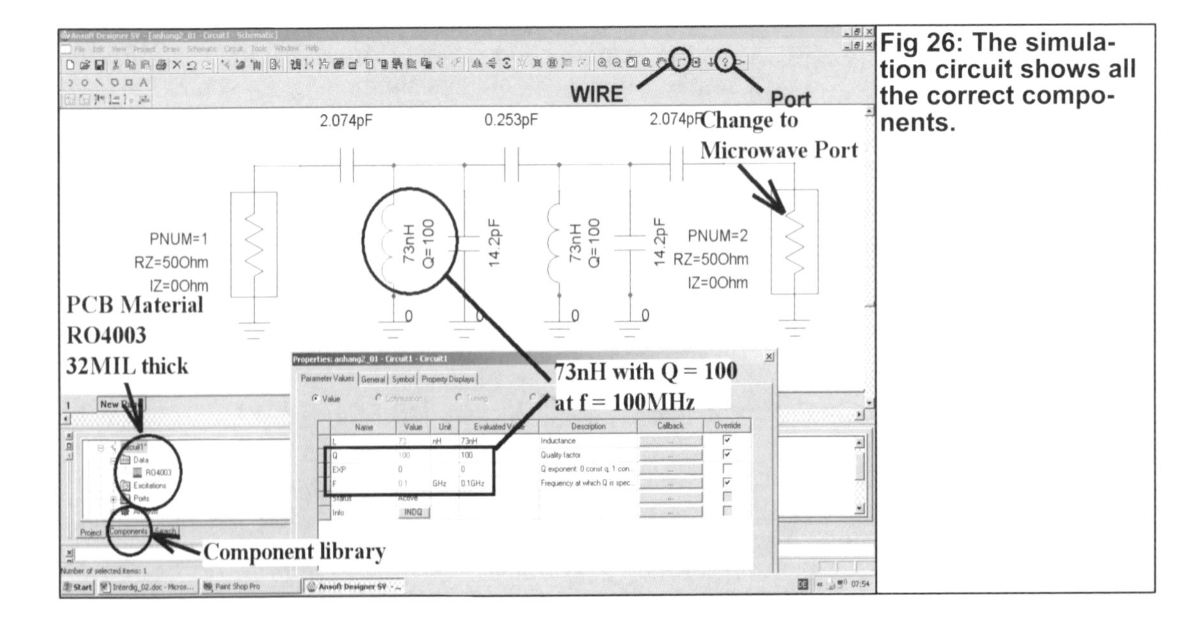

Fig 26: The simulation circuit shows all the correct components.

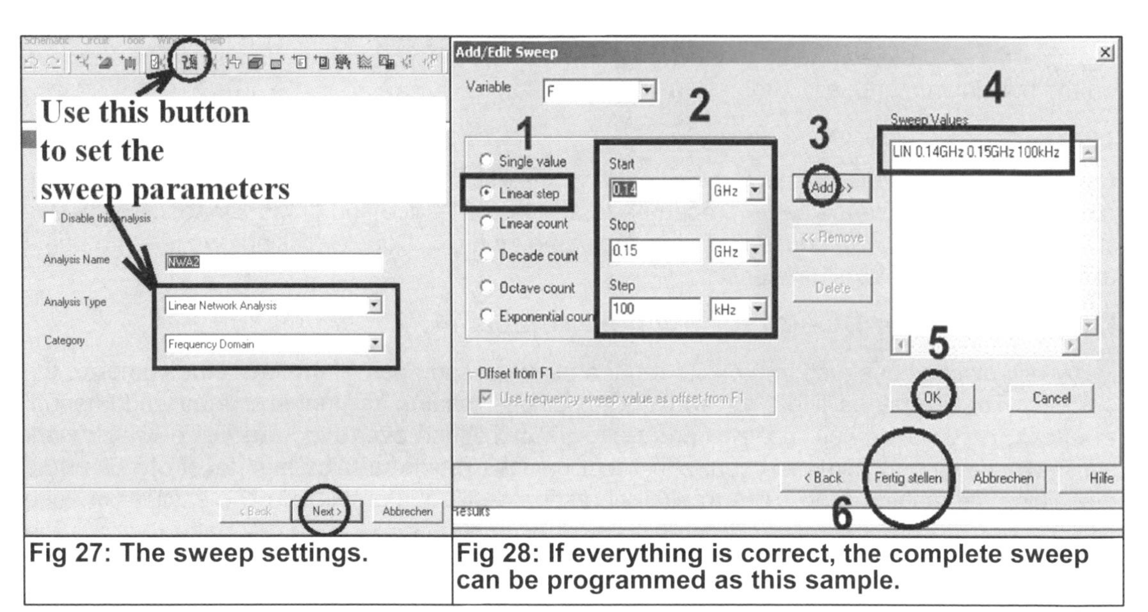

Fig 27: The sweep settings.

Fig 28: If everything is correct, the complete sweep can be programmed as this sample.

Step 2:

Select "Add" on the next menu to show the sweep programming (Fig 28) and set the following:

1: Control "linear Sweep"

2: Sweep attributes (140MHz to 150MHz in 100kHz steps)

3: Press "Add"

4: Check the sweep values selected

5: Press "OK"

6: Lock the sweep programming with "Finish"

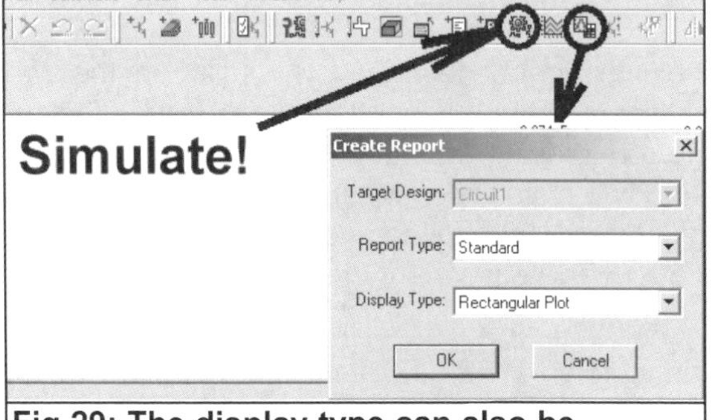

Fig 29: The display type can also be Smithchart. Also different forms of representation can be selected.

Step 3:

Pressing the simulate button starts the simulation but nothing is displayed until the create

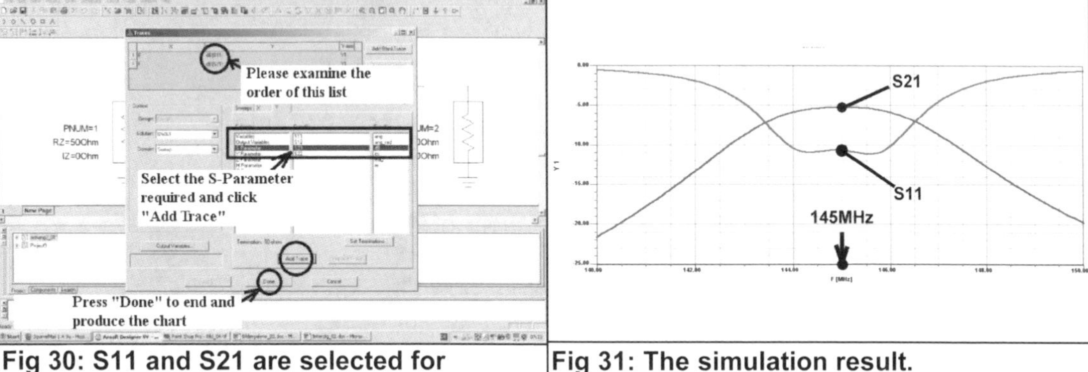

Fig 30: S11 and S21 are selected for presentation.

Fig 31: The simulation result.

report button is pressed (Fig 29). Check that "Rectangular Plot" is selected; this can be changed on the pull down menu, e.g. Smith Chart representation.

Step 4:

Use the "Traces" menu to add S parameters to the list shown in Fig 30. Select S11 and then click "Add Trace". Use the same procedure to add an S21 trace and press "Done". The display shown in Fig 31 is now produced which is the same as Fig 8. Double clicking on the appropriate axis can change the axis divisions.

Band Pass Filter Design for μWave Projects [4]

Many microwave construction projects require narrow band pass filters for various parts of the system. These may be filters for frequency multiplier chains and receiver front end filtering. Whereas there have been many different designs published over the years, all of which work satisfactorily when constructed correctly, most designs require a fairly high level of skill in the machining and construction. The traditional resonator filters are split into two similar but quite different design concepts. The "Comb-Line" filter is named because the resonator rods are like the teeth of a comb and are all grounded at the same end. By contrast the "Inter-Digital" filter transposes the grounded end for every other rod. For both of these filter types the spacing between adjacent resonator rods or "fingers", lengths and diameter of the rods and the tapping up from the grounded end for the input and output connectors is critical. Any errors can result in poor impedance matching, excessive ripple in the pass band, excessive insertion loss and poor attenuation in the stop bands.

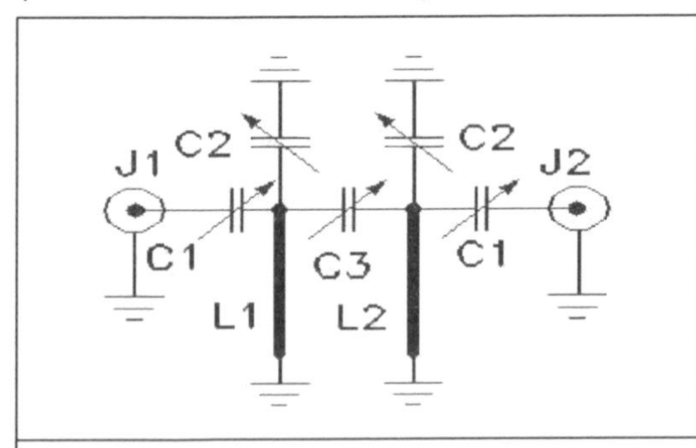

Fig 32: Equivalent electrical circuit of the band pass filter.

The filter to be described is similar to the two traditional filters but avoids the problems of accurate rod spacing and length. This is achieved by using two critically coupled cavity resonators with adjustable top coupling and impedance matching to 50Ω. The equivalent electrical circuit is shown in Fig 32.

Connectors J1 & J2 are SMA panel mount connectors with a long stub nose, which have the extended centre pin surrounded by PTFE. These are made by several manufacturers and are designed to allow the SMA bulkhead connector to be bolted to a wall of considerable thickness and the hole bored for the centre pin forms a 50Ω coaxial transmission line with the PTFE as the dielectric. These are available in various standard lengths; the longest being 0.59" (15mm) and the shortest is 0.16" (4mm). The centre pin protrudes a few millimeters proud of the PTFE. The diameter of the PTFE dielectric is nominally 4.1mm and a hole of 4.2mm bored in the equipment wall allows a push fit, with constant 50Ω impedance. A typical SMA stub nose connector is shown in Fig 33. They are available with male or female ends to suit the application.

Fig 33: SMA connector with long centre pin and PTFE dielectric.

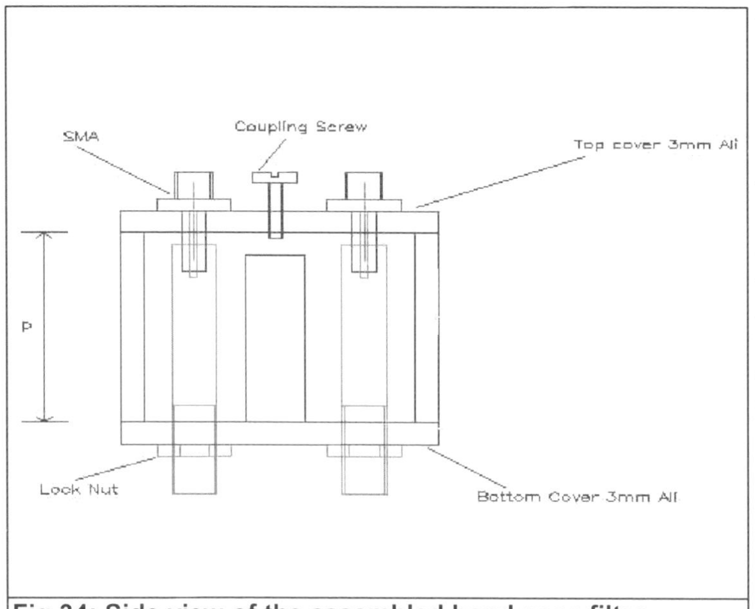

Fig 34: Side view of the assembled band pass filter.

Tuning capacitors C1 & C2 are formed by the SMA centre pin and the resonator rod in a "coaxial capacitor". Capacitor C3 is the top coupling capacitor and is formed by the fringing capacity between the top of the first resonator rod and the second resonator rod, each in their own shielded cavity. This capacitor needs to be an extremely low value, typically less than 0.05pf. To obtain the very small value of capacity required the two resonator rods to "see" one another by a small iris cutout between the adjacent cavities. The aperture of the iris can be controlled by the coupling screw, which partially blocks the iris. This is shown in the general arrangement drawing, which is a side view of the filter. Fig 34 shows the mechanical assembly of the 2-resonator filter.

The body of the filter is made from a block of aluminium 30mm x 50mm in profile and the height of the block (P) is determined by the operating frequency. The exact dimensions of the outer surfaces are not critical. Although 30mm x 50mm is quoted the piece of aluminium used by the writer is actually 31.75mm x 50.8mm, this being an imperial size of 1.25" x 2". Most metal wholesalers sell metal by weight and it is not uncommon to find that sizes are rounded up or down to the nearest whole number in the metric measurement system. This is because the cost of a new extrusion die to form the various sections is very expensive and rather than make a new die to suit the metric measurement system they round the size when converted to metric dimensions. Often when a piece of metal bar or sheet is measured it turns out it is in fact an imperial size in disguise!

The filter was simulated using the ARRL Radio Designer software (Compact Software sub-set) and the various dimensions were determined for popular frequencies from 1152MHz up to 2304MHz. These are detailed in Table 1. The dimension P in Fig 34 will be defined later.

Please bear in mind these dimensions were obtained from computer simulations, some "tinkering" may be required to obtain the correct response for a particular filter. Details of corrective measures are contained in the sections "Tuning Up" and "Correcting Errors".

The top and bottom covers are made from aluminium sheet of 3mm thickness and contain various holes. The resonator rods are also aluminium but an unusual diameter. They are 9.525mm in diameter. 9.525mm, (which we will round to 9.5mm for simplicity), is in fact the imperial dimension of 3/8" (0.375 in) and the reason was the need to thread the bottom ends of the resonator rods with a suitable fine thread. There is no reason you cannot use metric bar stock of 10mm but fine pitch metric taps and dies are more difficult to obtain and more expensive. Standard (coarse pitch) metric threads are too coarse and "wobbly" for precision work such as this application.

The writer has a 3/8" x 32 UNEF set of taps and dies for threading material. 3/8" x 32 UNEF is the thread used for BNC connectors, in fact all the popular RF connectors traditionally used are based on imperial thread types. For example, N type connectors use 5/8" x 24 UNEF, SMA connectors use ¼" x 36 UNS and TNC connectors use 7/16" x 28 UNEF. The numbers 24, 28, 32 & 36 etc is the number of threads per inch. As many of these popular connectors are also specified under the NATO stock number system (NSN) they are not going to change any time in the foreseeable future!

A fact that the "Euro-crats" almost got horribly wrong recently when it was proposed to ban the use of imperial measurements or thread sizes. Fortunately common sense prevailed and the imperial thread and measurement system has been granted a new life.

Note: UNEF stands for Unified National Extra Fine, an American standard thread type, and is a finer pitch than normal UNF threads. The threads per inch for the 3/8" UNF thread is 24. 32 threads per inch are equal to a pitch of 0.03125 in or 0.79375 mm. A standard M10 thread has a pitch of 1.5mm and the Metric Fine pitch is 1mm. Some metric sizes are available in 3 different pitches. For example, M9 is available in 1.25mm, 1.00mm and 0.75mm, these being the standard, fine and extra fine versions.

Simulated response

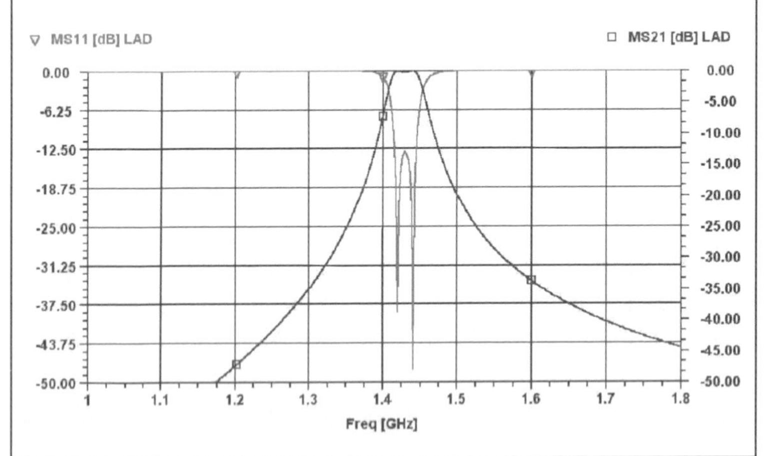

Fig 35: Simulated response plot for the 2-resonator filter at 1420MHz.

The response of the 2-resonator filter for 1420MHz for a Hydrogen Line radio astronomy receiver is shown in Fig 35. You will note that the top coupling tends to make the filter behave like a high pass filter on the higher frequency part of the response plot. Normally this is of no concern as long as the attenuation obtained is adequate for the intended purpose. The ripple in the pass band can be adjusted to close to zero. In Fig 35 it is only 0.1dB maximum.

Note also the excellent impedance matching possible. The value of the input reflection coefficient (VSWR) MS11 is typically –30dB or better for all the filters simulated. The stop band attenuation is comparable to a 3 or 4 resonator comb line or inter digital filter but the insertion loss is much lower because more resonators equate to more loss. This means that whereas we have previously needed to use 3 or 4 resonator filters to get the required stop band attenuation we no longer have to with this new design. It also makes the filter less expensive and smaller for the same performance.

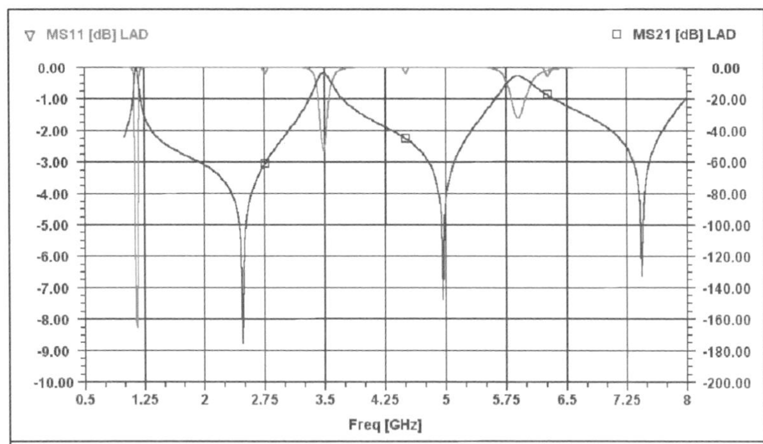

Fig 36: Wide plot of the filter response.

Some facts about resonator filters.

Many amateurs are not aware of the facts about resonator filters, like the comb line and inter digital filters. Many assume they are "monotonic", that is, they only respond to a narrow band centred at the operating frequency. Unfortunately this is not the case. All resonator filters have spurious responses on higher frequencies when the line appears as 3λ/4, 5λ/4 etc, which are not directly harmonically related, although they are close to the odd order harmonics. The reason they are not exact odd order multiples is because the fringing capacity, and any loading capacity, alter the effective electrical length as the frequency rises. Fig 36 shows a typical comb line, inter-digital or this new design filter across a wide range of spectrum. This filter is one designed for 1152MHz and it has spurious responses at about 3.5GHz and 5.8GHz as well as many higher frequencies.

Hence, these types of filter are unsuitable for eliminating 3rd harmonic energy in a transmitter feeding an antenna. Also, because of the very high Q and hence impedance at the top of the resonator rods, the RF voltage may becomes excessive and may cause the filter rods to arc to ground with as little as 1W of power. So these types of filter are best used where the power levels are low. Harmonic filtering is best obtained with low pass filters such as the "Rod-Bead" filter for the wave spectrum.

The other factor is the length of the resonator rod versus the impedance. The coaxial resonator is the easiest one to understand. The characteristic impedance, (surge impedance), of a coaxial line is determined by the physical dimensions and the dielectric material. We often use air as the dielectric (as in this case) and so the dielectric constant is 1.00. PTFE has a dielectric constant of 2.12 greater than air.

As the characteristic impedance is varied the physical length of the resonator rod also needs to be varied to maintain resonance. For a low impedance resonator the length needs to be increased to maintain resonance, conversely for a high impedance resonator the length needs to be reduced.

In coaxial resonators approaching a quarter wavelength electrical length the optimum impedance is 70.7Ω. But this is only true where the end capacity is simply that naturally existing, the so-called "fringing capacity". For a capacitively loaded, and hence shortened, quarter wave resonator this no longer holds true and we can choose any impedance within reason. In this design the impedance was chosen purely based on mechanical constraints. The resonator rod was chosen to be 9.5mm (3/8") because of the availability of suitable threading tools. The simulations showed no significant deterioration as the impedance was varied over a

Table 1: Dimensions of resonator rod length.

Frequency (MHz)	Length of resonator (mm
1152	60.4
1268	55.9
1296	53.5
1420	48.5
1572	43.75
2160	31.8
2304	29.1

wide range. Lower impedance made the resonator rods longer but the insertion loss in the pass band tended to be slightly less than higher impedance rods. Practicality must enter the argument eventually and the diameter of the resonator bore was chosen to be 17mm simply because a suitable drill was to hand to drill the filter block. Using the standard formula for air dielectric coaxial lines:

$$Zo = 138 \log (D/d)$$

Where:

d is the resonator rod diameter

D is the bore diameter

This gave a value of 34.71Ω with the dimensions chosen. The ratio of the two diameters is 1.78:1, as long as we stay close to this ratio then little change will be required in the dimensions of the resonator rod lengths. The software simulated the required capacitor values and the physical length of the resonator rods for the various operating frequencies. These are listed in Table 1.

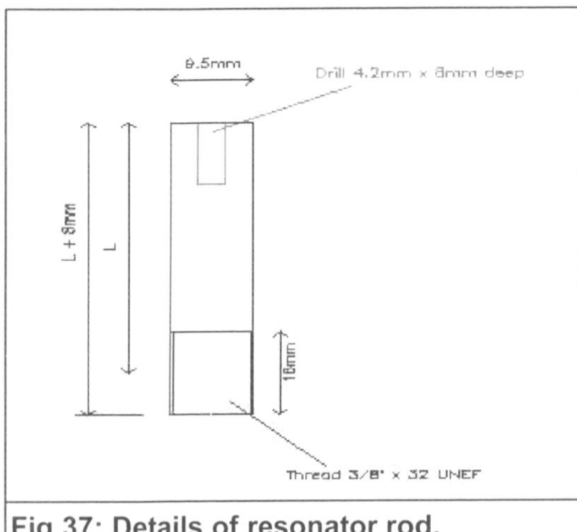

Fig 37: Details of resonator rod.

The length quoted in Table 1 is the length for resonance; this is not the total length of the resonator rod. This needs to be calculated from the information contained in Fig 37. The rod is threaded for a length of 16mm and the dimension from the bottom cover to the top of the rod is the dimension quoted in Table 1. This is denoted by dimension L in Fig 37.

In Fig 37 the total length of the resonator rod is the table dimension plus 8mm.

The dimension P shown in Fig 34 can now be stated. This dimension is the resonator effective length plus 6mm. For example, for 1420MHz the effective length of the rod is 48.5mm and the dimension P is hence 54.5mm, being the internal height of the cavity. This is the height of the filter block required and it should be machined to this dimension with a smooth surface finish on the top and bottom faces. An error of 0.5mm either way is not going to make the filter not work; it simply will not work as well as it could!

Fig 38 & 39 give details of the top and bottom covers and Fig 40 gives details of the filter block.

Construction method

For this you will require certain items. At the minimum a steel ruler marked in millimeters, a marking scriber, small centre punch, small hammer, right angle square, and a selection of drills and taps. A vernier calliper if available is a good substitute for the ruler and scriber and the lines can be scribed with the tips of the vernier jaws once set to the required dimension.

Note: *Although the preferred screws are called up as metric types there are still a large number*

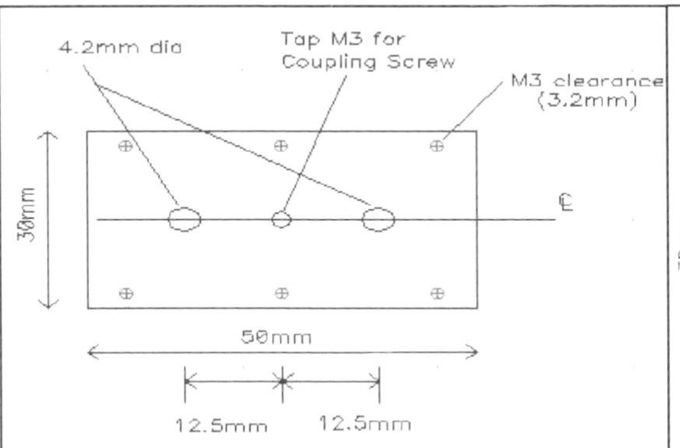

Fig 38: Top cover details. Material 3mm aluminium sheet.

Fig 39: Bottom cover details. Material 3mm aluminium sheet.

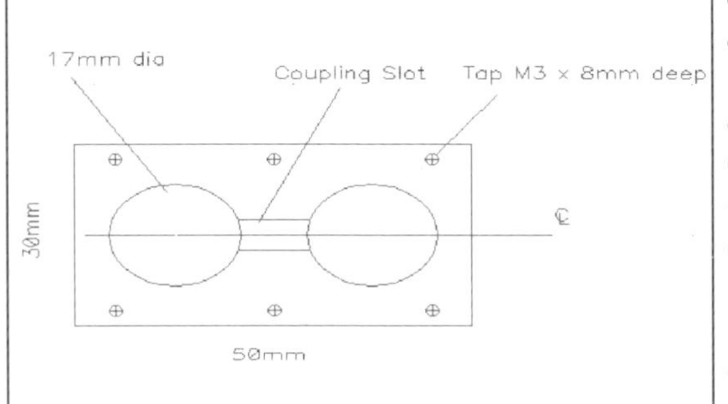

Fig 40: Filter block details, top view.

Fig 41: Top and bottom covers cut to size and top cover marked out and centre punched for drilling. The bores have also been marked to ensure the cover fixing holes do not intrude on the bores.

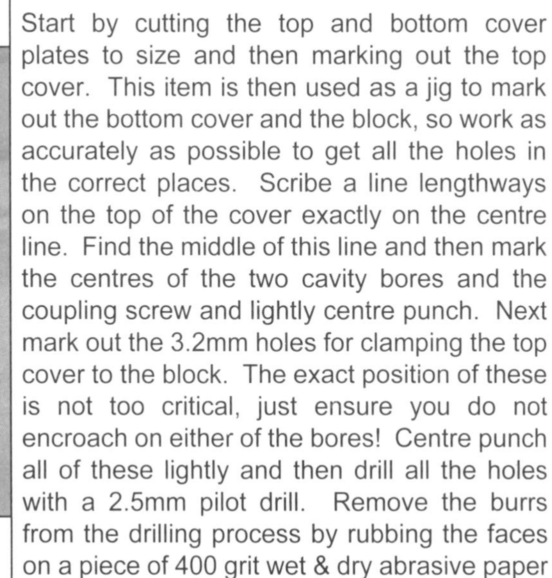

of amateurs who have British Association taps and dies and a variety of BA screws and nuts. The nearest BA size to the M3 screw is 6BA and for the M2.5 it is 8BA. BA threads are still extensively used for very small screws, especially in spectacle frames, whereas the metric sizes stop at 2mm. The smallest BA screw is 25BA, which has a diameter of approximately 0.25mm. BA screws where in fact the first available thread system based on metric dimensions and date back to the early 1850s. 0BA is the largest and is 6mm in diameter.

Start by cutting the top and bottom cover plates to size and then marking out the top cover. This item is then used as a jig to mark out the bottom cover and the block, so work as accurately as possible to get all the holes in the correct places. Scribe a line lengthways on the top of the cover exactly on the centre line. Find the middle of this line and then mark the centres of the two cavity bores and the coupling screw and lightly centre punch. Next mark out the 3.2mm holes for clamping the top cover to the block. The exact position of these is not too critical, just ensure you do not encroach on either of the bores! Centre punch all of these lightly and then drill all the holes with a 2.5mm pilot drill. Remove the burrs from the drilling process by rubbing the faces on a piece of 400 grit wet & dry abrasive paper resting on a smooth flat surface.

Clamp the bottom and top cover together and

Fig 42: Using the drilled top cover as a jig to drill the bottom cover. Note the thin piece of wood placed between the cover plate and the machine vice to prevent damage when the drill breaks through the bottom cover

Fig 43: Using the top cover to spot through onto the block for the various holes.

Fig 44: Using a tap wrench to start the threads guided by the tap held in the drill chuck.

make witness marks with the 2.5mm drill on the bottom cover. Drill the bottom cover with 3.2mm clearance holes for the M3 fixing screws. Note that the coupling screw hole is required on the bottom cover as an additional fixing to clamp the plate between the bores.

Clamp the top cover to the undrilled block, line it up exactly and spot through all the pilot holes with the 2.5mm drill to form witness marks for the drilling. Place the block to one side and continue with the top and bottom covers. It is a good idea to mark the parts in some way so you don't assemble them incorrectly. A light centre punch mark on one end of the top cover and a corresponding one on the block is one way. Whichever method you choose ensure it will not be removed in subsequent finishing operations.

Open up the 2.5mm pilot holes in the top cover to either 3.2mm or 4.2mm as appropriate. Tap the centre hole for the coupling screw M3. The tap must go through exactly perpendicular to the surface. The best way to achieve this is to hold the tap loosely in the pillar drill chuck with the cover resting on the drilling machine table. Clamp a tap wrench onto the plain portion of the tap and use this to rotate the tap. Apply some tapping lubricant. After two or three revolutions you can finish the tapping by hand with the tap wrench fitted to the square end of the tap. Deburr both sides and give the surface that will clamp to the block a rub on some 400 grit wet & dry abrasive paper on a smooth flat surface.

Treat the bottom cover in the same way and open up the 2.5mm pilot holes to either 3.2mm or 8.5mm tapping size for the 3/8" threaded holes. Work up to the 8.5mm size in small steps so the hole is truly round and in the correct place.

Now we can drill the block. For this you will need a selection of drills varying in size up to 17mm final size. Clamp the block in a machine vice and clamp the machine vice onto the table of the pillar drill.

Do not even consider trying to hold the machine vice by hand. It will be ripped out of your hands under the cutting force and most likely cause a serious injury and damage to the block. Even if you get away without any injury the bore will not be straight and perpendicular to the top and bottom faces. Ensure the block is square and well tapped down before tightening the machine vice jaws.

Start with a small diameter drill (4.2mm) and drill as far into the block as possible. Standard

Fig 45: Filter block machined to size ready for drilling and two resonator rods with slotted ends and lock nuts.

Fig 46: Checking the block is vertical in the machine vice with a set square.

Fig 47: 6mm pilot holes drilled in block.

length jobber drills will probably be too short for the lower frequency filters. Change to a 6mm drill and run the drill slowly, not more than 200rpm. Pour on an ample supply of cutting lubricant, I use Tapmatic #2™ for aluminium but light machine oil is also OK.

When the drill exits the bottom of the block ensure it doesn't drill into the machine vice! Position the block so the bore is in the centre of the machine vice over the hole. If you move the block during the drilling operation ensure it is tapped down square with a small hammer and piece of wood to protect the surface. A set square placed on the drilling machine table and touching the side of the block will show if the block is truly vertical.

Work up to the final hole size in small increments. We need the hole to be straight and a high surface finish. Drill out to about 16mm. When about to use the final drill, apply plenty of lubricant and run the drill as slowly as possible, for 17mm about 80 to 100rpm is about correct. Do not force the cut, let the drill work at its own pace, on the other hand do not let the drill rub, it must cut to get a good surface finish. Remove the drill bit every few millimeters of depth and clear away the swarf and apply more cutting lubricant. The better the bore finish the lower the insertion loss in the finished filter. If the bore turns out a bit ragged then you can polish it with some abrasive paper wrapped around a piece of wooden dowel. Alternatively an automotive brake wheel cylinder hone in the pillar drill will give a mirror like finish. If you have an adjustable reamer then this will also give a high surface finish to the bore. If planning to use a reamer then leave the hole slightly under-size and remove the last 0.2mm with the reamer to bring to finished size.

Fig 48: 17mm holes drilled in block.

Having drilled both bores turn your attention to the fixing screws. Drill the cover fixing screws (using the witness marks made earlier) 2.5mm holes about 8mm deep and tap M3. Again use plenty of lubricant, small drills and taps break easily if used without cutting lubricant. When drilling the small holes into the block remove the drill after 2mm to clear the swarf and apply more cutting lubricant, repeat until the full depth is reached. A small paint brush dipped in cutting oil will remove the swarf and apply new cutting lubricant.

Turn the block upside down and repeat for the bottom cover holes. Remove any burring from around the bore face, a

Fig 49: Coupling slot machined into top of block.

penknife is a good tool for this as aluminium is soft and easily removed.

Now the coupling iris slot needs to be machined. This needs a milling cutter 4mm in diameter. For this operation a 2 or 3 flute slot drill is the best tool. You will probably have to get this part performed by someone who has the correct tooling if you do not have access to suitable machinery. A model engineering society nearby will be the best option. Asking for assistance is not a big thing! Machine the slot 5mm deep between the two bores.

Finally rub the two block faces on the 400 grit wet & dry abrasive on a smooth flat surface to remove any burrs.

Resonator rods

The resonator rods are a simple threading operation; again a model engineering society will probably undertake this part if you do not have the correct machinery. It is important the threads are well formed and square, do not attempt this by hand! To get well fitting threads requires the threading die to be opened up slightly so the threads are slightly oversize. A lathe with a tail stock die-holder is the best method as this ensures the threads are true. Apply plenty of cutting lubricant to prevent torn threads in soft aluminium.

For those without access to a lathe the job can be done in the pillar drill with care. Place the die holder with the die inserted flat on the drilling machine table so it is centralised with the center hole. Clamp the die holder down. Place the piece of rod to be threaded into the drilling machine chuck with sufficient extending for the length of thread. Bring the chuck down so the rod starts to enter the die and rotate the chuck by hand. Do not attempt this with the drill switched on! Work the drill chuck forwards and backwards to clear the swarf and apply plenty of cutting lubricant. After about 5mm of thread have been formed the rod with the die-holder attached can be removed and clamped in a bench vice and the threading completed by hand with the die holder.

Having threaded the resonator rod, the rod can be cut or parted off to the correct length and the end faced off square. If no lathe is available then a flat file placed on the drilling machine table and the rod held in the drill chuck will do almost as good a job. Ensure you do not damage the threads with the chuck jaws. Bring the rod in contact with the file with the drill running slowly.

Fig 50: Threading a resonator rod in a lathe with a tail-stock die holder. The rear tool-post contains a parting off tool as a gauge for how far the threads are cut.

A centre drill in the lathe tailstock drill chuck is used to make a pilot hole for the 4.2mm diameter tuning capacitor hole. Drill this 8mm deep and slightly chamfer the corner of the rod with a fine file and the entry to the hole with a centre drill. Finally, polish the rod on the plain portion to a high finish. The writer uses a piece of "Scotchbrite" pad ™ made by 3M, which gives a mirror finish. The same material is found on kitchen scouring pads with the pad bonded to a piece of sponge. Apply mineral turpentine as a lubricant. Alternatively a strip of 400 grit wet and dry abrasive paper applied to the rotating rod will produce a high surface finish.

Fig 51: Completed filter.

Lock nuts can be used from BNC single hole mount types, or many early potentiometers of British manufacture also used 3/8" x 32 UNEF thin nuts and some still do. If you have access to a lathe it is a simple machining operation to make two lock nuts from 13mm hexagonal bar. If you wish you can make a saw cut across the bottom of the rods to allow a screwdriver blade to be inserted for easier adjustment. Put two hacksaw blades in parallel to form a wider slot.

Assembling the filter

Clean all the parts with alcohol or similar cleaning spirits to remove the bulk of the cutting lubricant and wash in warm soapy water (dish washing liquid) to remove any greasy finger marks. An option is to have the parts silver plated for best results; the writer hasn't had the opportunity to treat his filters this way at present.

Begin by screwing the two resonator rods into the bottom cover and fix the lock nuts loosely. Measure the distance from the bottom cover to the top of the rods and adjust to the dimension given in Table 1. Tighten the lock nuts.

Place the bottom cover on the bottom of the block and fix in place with M3 x 10mm cheese head screws, preferably stainless steel hardware. Temporarily fasten the top cover on the top of the block with two M3 screws positioned diagonally and place the two SMA connectors into the 4.2mm clearance holes for the stub nose portion. If your workmanship is OK then the centre pins should enter the hole in the rod freely. Should one or both SMA pins be out of alignment with the resonator rod, then open up the 4.2mm hole to 4.5mm and try again. If it all aligns correctly then mark the fixing holes for the SMA connectors by spotting through the connector flange with the 2.5mm drill and drill (2.1mm drill) and tap for M2.5 screws. Deburr the top cover and fix in place with M3 screws. Fasten the SMA connectors with M2.5 x 6mm screws. Ensure the ends of the screws do not protrude more than 2mm into the cavity, if too long then file them down to nearly flush with the bottom of the top cover. Place an M3 x 12mm stainless steel screw into the coupling screw tapped hole with a lock nut. Screw the coupling screw in until it touches the bottom of the iris slot and tighten the lock nut.

Tuning Up the filter

For this we need some means of generating a low power signal that can be varied in frequency and a means of measuring the power exiting the filter. How you do it will largely depend on the type of test equipment available. A μwave signal generator and milliwatt meter such as a HP-432 is one option. The best instrument is a Vector Network Analyser, if you can get access to one for the adjustments. This will display not only insertion loss, stop band attenuation but also input and output matching for a full 2-port measurement system. Unfortunately not all amateurs have access to such luxuries!

Assuming you have the bare minimum equipment, then apply a low power signal at the centre frequency and measure the output power at the other SMA connector. Start by reducing the depth of the coupling screw. If you are lucky then the power should rise up to a maximum. Now try adjusting the resonator rod lengths by small increments. Loosen the lock nuts a little and rotate the rod either more in or more out. The power should peak some more. Measuring the power from the signal generator with the filter removed and noting the reading can make an estimate of the insertion loss. By inserting the filter the measured power will drop, by, say 1dB.

Now alter the signal generator frequency either side of the filter centre frequency. The power indicated should rise to a peak as the generator comes “on-frequency” and drop off as we move away from the centre frequency. As we sweep across the filter centre frequency we may see a bit of ripple in the pass band as the output power varies. This is a pointer to too much top coupling as the response plot now has a big dip in the centre of the curve. The two resonator rods being slightly different in length and so resonating on different frequencies can also cause it.

Correcting Errors

So far we assumed everything was working OK. But what about the case where we have a problem? We have quite a bit of latitude to solve minor errors because we have so much which can be adjusted.

Suppose the insertion loss is excessive but the response is OK. This can be caused by several things, poor mechanical joints, which cause high resistance, and also insufficient top coupling etc. Also the input and output matching have a direct bearing on the filter ripple and insertion loss. The resonator rod as it is increased in length does two things. Firstly, it moves the resonant frequency lower, but is also increases the input/output coupling capacity as the PTFE stub is inserted further into the rod. To decrease the effect of the coupling capacity being excessive we have several choices.

Firstly, we can shorten the centre pin and PTFE stub. It is a good idea to cut the exposed centre pin, which stands proud of the PTFE, so that it is flush. This reduces the coupling a little. Secondly we can reduce the length of the PTFE stub that is able to be inserted into the rod. This is a bit drastic, because once we have cut it off we can’t glue it back on again! So make a small square of aluminium 1.6 to 2mm thick the same dimensions as the SMA flange to space the SMA connector away from the top cover. Drill clearance holes for the 2.5mm screws (2.7mm) and the centre pin of 4.2mm. If this improves the response then we can safely trim the PTFE and pin back.

Another method is to open up the 4.2mm hole in the end of the resonator rod to decrease the coupling capacity and the rate at which it changes. Try drilling this out to 4.5mm and see the effect it has.

Insufficient top coupling, if the coupling screw is fully out, this can be corrected by machining the iris slot a little deeper or wider. It is preferable to make it wider than deeper. The reason is that if we find the coupling is now excessive we can drill out the M3 coupling screw hole on the top cover to 3.3mm and tap for a M4 screw. If the slot is too deep than a shim of a sliver of aluminium can be cut and glued into the iris slot with a drop of "Superglue" to reduce the depth. Alternatively take a small amount off the top of the block, about 1mm, and try again. Once the first filter is working satisfactorily dismantle it and note all the critical dimensions down on paper should you wish to make more filters for this frequency.

A Low Loss 13cm High Power 90° Hybrid Combiner [5]

With the availability of cheap surplus CDMA power amplifiers, producing up to 200W on 13cm when properly tuned, a solution was searched for combining two or more amplifiers in order to generate a serious power level, preferably for EME communication.

Looking around on the professional microwave components market for suitable combiner units revealed only very few devices, which could safely handle 500+ watts of combined power. Most of these very expensive combiners do not reach full specification data in the 13cm amateur band, as their design frequencies are somewhat different from amateur needs. Thus, they typically provide isolation on 2.3GHz of 15 - 20dB, which is of course acceptable, but not top-notch performance.

A viable solution for a do-it-yourself unit could be the well-known 3dB quadrature branchline coupler, rendering a reasonable isolation between the two sources at an adequate broadband performance. Meanwhile, there are a lot of pcb designs available, some etched on low-loss PTFE substrate, but even with this material the dielectric losses confine the safe power level to 100W maximum.

A branchline coupler for 1.3GHz has been published by R. Bertelmeier, DJ9BV (sk), in 1994 [6], exhibiting almost perfectly symmetric coupling at very good isolation of >25dB for the source ports, ideal for coupling modern solid state amplifiers. It uses air striplines and the mechanical construction is such that it does not necessarily need a milling machine, the average ham with a jigsaw and some patience can achieve it. The dielectric losses are so low that the transferable RF power is mainly confined by the coaxial connectors used; up to 1kW can be handled, using precision N connectors. Special care was taken with the impedance match to the 50Ω connectors by a tapered pattern of the branchline terminations.

The idea now was to scale the dimensions of this 23cm hybrid to the 13cm band. The lenghth of the 4 quarter wave branchlines can be easily scaled to 2.3GHz, as can be the width of the striplines in order to match 35 and 50Ω, respectively. Unfortunately, simple numerical downscaling is not possible, as the DJ9BV coupler is a triplate structure, which means the impedance of the striplines is not only determined by their mechanical width, but also by the two groundplanes they see in form of the lid and the bottom of the enclosure of the coupler. To get an idea about the necessary clearance to the two ground planes (we are using a non-offset branchline configuration with symmetrical distances to the two groundplanes) and the resulting impact on the impedance of the striplines the following formula may be applied:

$$Z_0 = (60 / \sqrt{\varepsilon_r}) \ln(4h / 0.67 \pi W (0.8 + t))$$

Where:

h = distance lid to bottom

W = width of stripline

t = thickness of stripline conductor

ε_r = dielectric constant of air

Still this term is only an approximation, as it is preferably valid for narrow striplines and small ground plane distances. The first prototype was constructed on this basis and surprisingly it functioned very well on the first shot with the dimensions given in Fig 52.

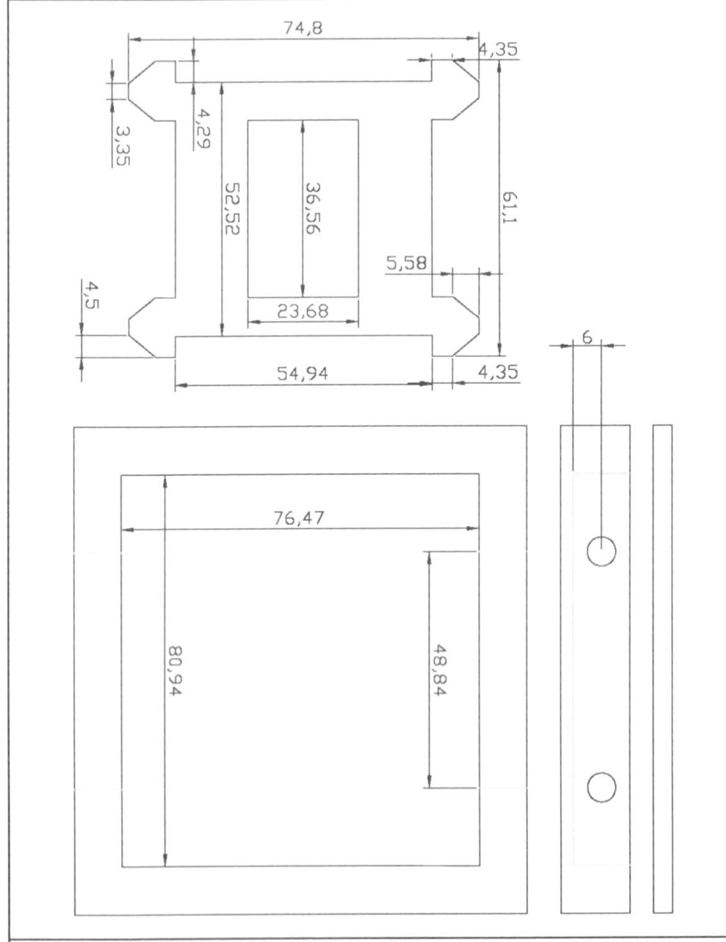

Fig 52: CAD drawing of the coupler courtesy ON7UN.

The branchline dimensions are critical, so the hybrid was cut out by a laser device from a 1mm thick sheet of copper, with the clearance in the casing being 12mm, resulting in a distance stripline to groundplane of 5.5mm. Special care has to be taken to keep this distance thoroughly at any location, when mounting the hybrid in the enclosure by soldering the tapered ports to the connectors, as this directly affects the 3dB coupling factor. Preferably the branchline hybrid is polished and galvanically silvered to reduce any skin effect losses, especially if the coupler is intended to carry high RF currents.

The enclosure is a milled aluminium case with a wall thickness of 5mm, so commercially available precision SMA connectors with 5mm long PTFE flange can be conveniently mounted in the walls. Although high-quality female SMA connectors may be used up to 400W at 2GHz, for the

Fig 53: The finished coupler with the cover removed.

Fig 54: The finished coupler.

sum port a female precision N connector or 7/16 connector is highly recommended. Figs 53 and 54 give an impression of the finished device.

The plots in Figs 55 - 58 give an overview of some essential electrical parameters of the 90° hybrid coupler as a broadband device, which was designed for an operating range from 2300 to 2400MHz:

The author has been using such a hybrid to successfully combine two identical high power amplifiers of 300W each, resulting in a combined output power of 600W on 2.3GHz without any excessive heat dissipation being noticed with any of the coupler's components. At 600W the unbalanced power measured at the isolation port, terminated with a suitable 50Ω resistive load, was only about 700mW. This amplifier stack has been used successfully e.g. during the OH0/DL1YMK moonbounce DXpedition in October 2009.

For any more detailed questions the author may be contacted [5]

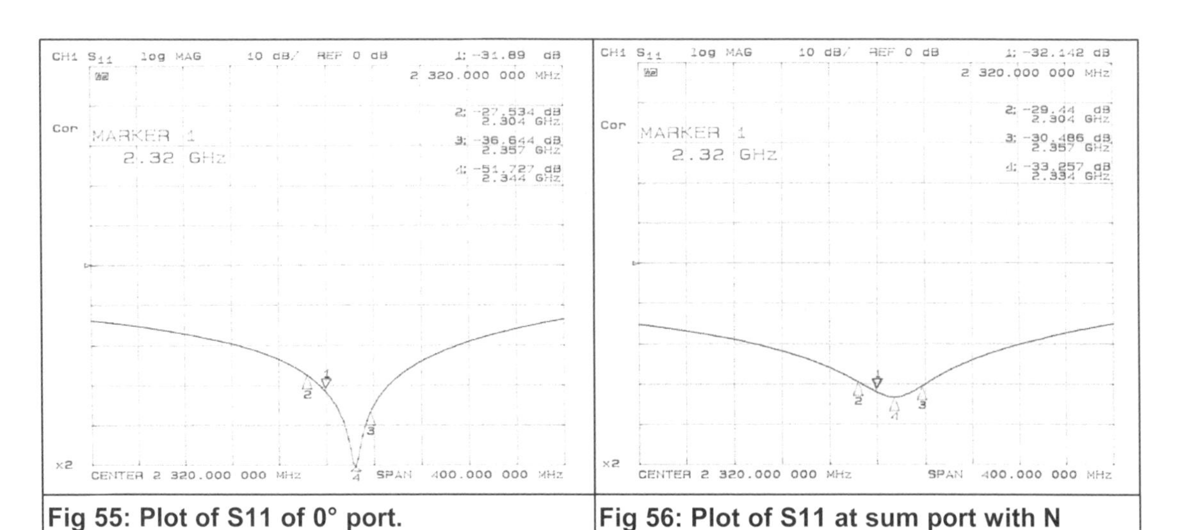

Fig 55: Plot of S11 of 0° port.

Fig 56: Plot of S11 at sum port with N connector.

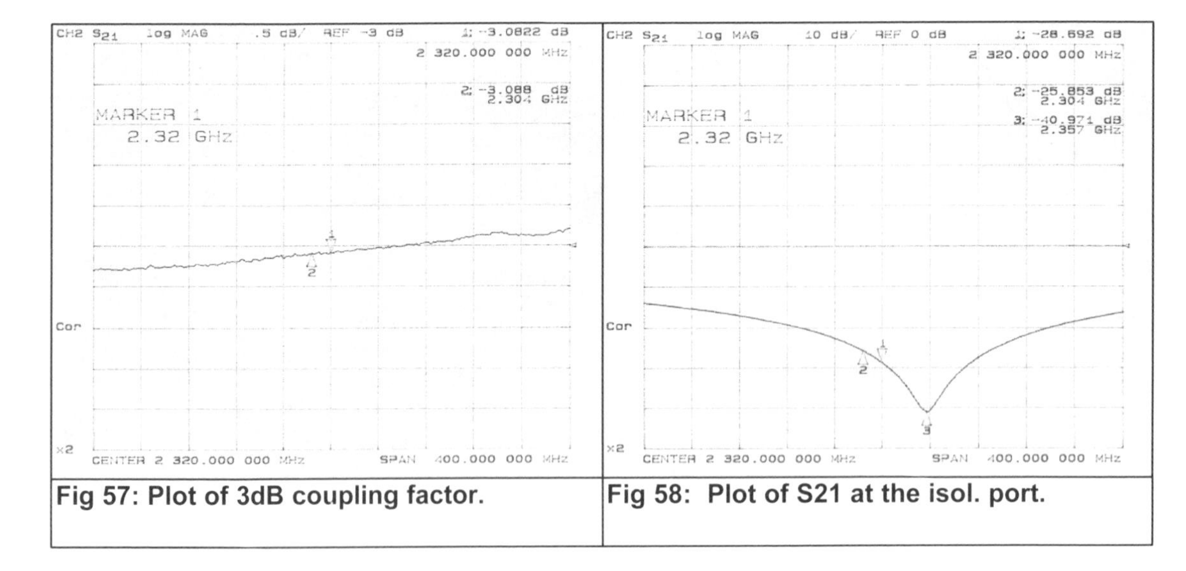

Fig 57: Plot of 3dB coupling factor.

Fig 58: Plot of S21 at the isol. port.

References

[1] Ansoft Designer SV project: Using microstrip interdigital capacitors, Gunthard Kraus, DG8GB, VHF Communications magazine 2/2009 pp 78 - 95

[2] www.elektronikschule.de/~krausg - is the main German web page and:

http://www.elektronikschule.de/~krausg/Ansoft%20Designer%20SV/English%20Tutorial%20Version/index_english.html - is the relevant English page

[3] Ansoft Designer SV (Student Version) - www.ansoft.com

[4] A New Band Pass Filter Design for μWave Projects, John Fielding, ZS5JF, VHF Communications Magazine 1/2008 pp 43 - 55

[5] Article written by Dr. Michael Kohla, DL1YMK – dl1ymk@aol.com

[6] R. Bertelmeier in: DUBUS communications, Volume 23, issue 4/1994, pp. 36 - 49

Chapter 5

Modifying Commercial Equipment

In this chapter :

- Eyal Gal 21.2 - 23.6GHz transceiver for 24GHz
- Converting the Ceragon 7GHz module for 5.7GHz
- DXR 700 TRV conversion to 5.7GHz
- Using the Eyal Gal 11GHz transverter on 10GHz
- Using the receive converter of White boxes on 24GHz

Until recently the only way to get active on the microwave bands was to build your own equipment but now there are a number of interesting items of commercial equipment that are coming on to the second hand market. Some of these can be used on the amateur bands without modification and some require modification to get them going on the amateur bands.

One of the most important things about using second hand equipment is to recognise it on a flea market table of on an Internet auction site. Having recognised that the piece of equipment is useful the next hurdle is to find out the specification and what all of the connectors are used for. One of the places to find all of that information is the UK Microwave group [1], they have a useful forum site where you can ask advice from other amateurs who have a wealth of knowledge. They also publish a monthly newsletter called Scatterpoint. The articles in this chapter have all appeared in Scatterpoint over the past two years and describe exactly how to use some of the commercial equipment, complete with pictures so that you can recognise the equipment when you spot it at the flea market.

Eyal Gal 21.2 - 23.6GHz Transceiver for 24GHz [2]

These units, shown in Fig 1, with the part number 6058-00 will work unmodified at 24.048GHz, with high side LO. This is good as it is impractical to change the RF circuitry. They consist of a combined transmit and receive system needing just an LO at half frequency, a Tx output filter and a couple of relays to make a complete 24GHz transverter. A block diagram of the unit is shown in Fig 2.

Measured performance on receive with a 432MHz IF, is as follows:

- Conversion gain: +26dB
- System noise figure: 0.3 to 3.9dB*
- Image rejection: (12.24GHz LO) –17 to -22dB*

Measured performance on transmit:

- Output: +30.5dBm at 1dB compression

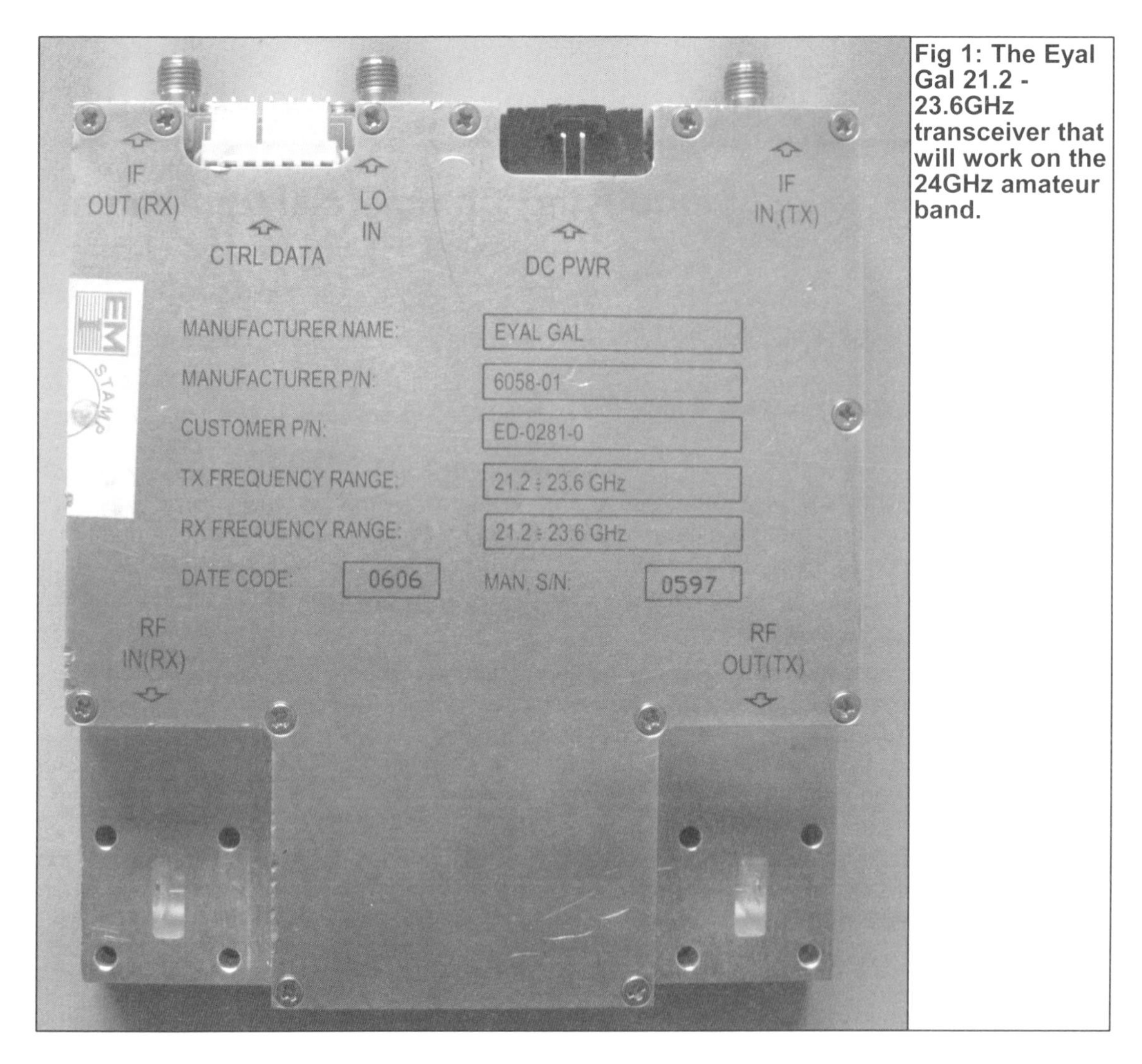

Fig 1: The Eyal Gal 21.2 - 23.6GHz transceiver that will work on the 24GHz amateur band.

- Gain: +53dB
- Saturated power output on transmit: >1.2W

Supply requirements:

	Receive (Tx Inhibited)	Full Output
+8V	600 - 700mA	0.9 - 1.0A*
+12V	10 - 25mA	500 - 750mA*
-12V	120mA	100mA

* The spread shows the results measured over three units

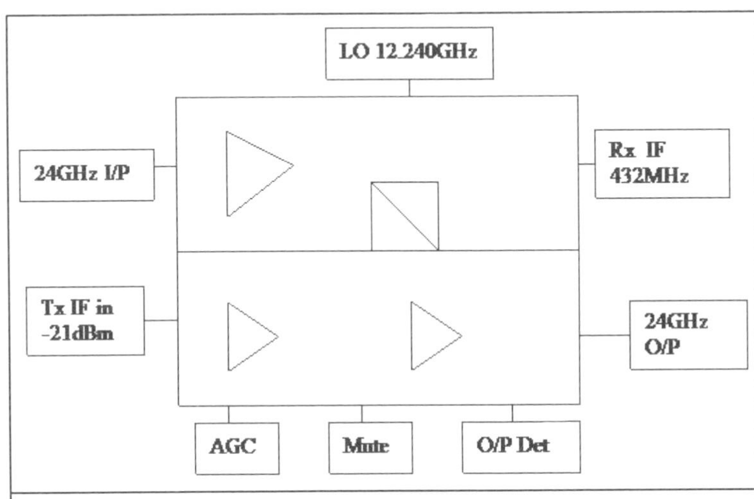

Fig 2: Block diagram of the Eyal Gal 21.2 - 23.6GHz transceiver.

My existing 24GHz system is an Alcatel unit using the Alcatel synthesiser, already with high side injection, and 10MHz reference. So I was ready to try the Eyal Gal block.

A specification sheet was available on the Internet [3]. This suggested that operation on 24GHz might be possible. The original IF was 2.8 – 4.1GHz, so would they work with a 432MHz IF? The IF response measured was essentially flat down to 100MHz, so it all looked hopeful. However, the RF response has rolled off around 4dB at 24.048GHz. However this still leaves around 26dB conversion gain.

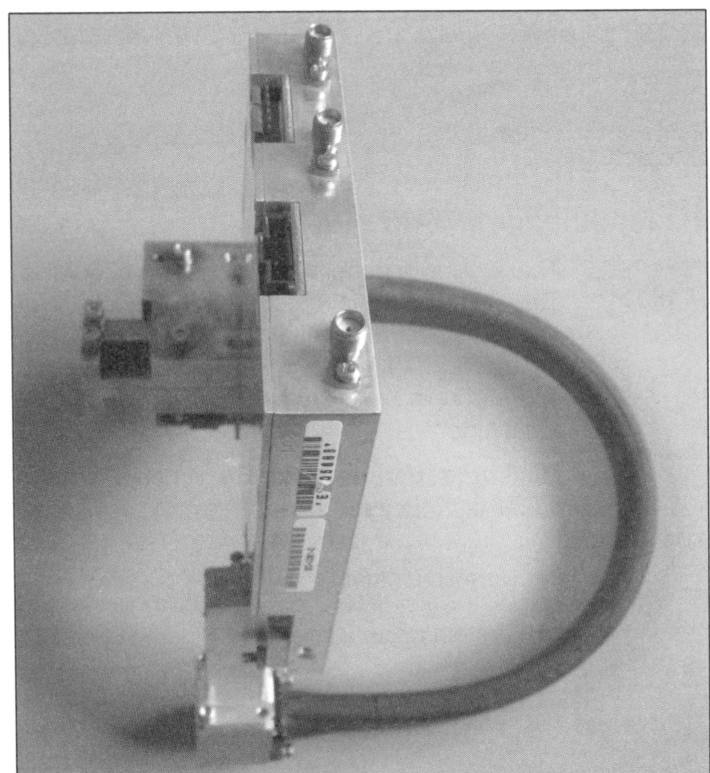

Fig 3: The authors Eyal Gal transceiver with filter and PIN waveguide switch.

The real novelty for me was that the available data included both pin-out, and voltages. The modules are designed for linear operation, so it was hoped to get around +27dBm out. In fact the first unit that I tried gave +31dBm, a very useful power level at 24GHz. Performance on receive is good with +26dB conversion gain, with the image -17 to -22dB and noise figure of 3.3 to 3.9dB. With the high transmit gain, it is necessary to use an input attenuator, and if required, to use the AGC control to turn the gain down.

On transmit it is necessary to use a filter on the output to reduce the LO and image to acceptable levels. My 24GHz unit with filter and PIN waveguide switch is shown in Fig 3. The block dissipates a fair amount of heat when in transmit mode, so additional heat sinking may be required for FM use.

Pin-out on the two connectors as shown in Fig 1, left to right, is as follows:

Connector 1 (6 way)

6	5	4	3	2	1
n/c	+8V	-12V	0V	n/c	+12V

Connector 2 (7 way)

7	6	5	4	3	2	1
Rx Agc	Tx Det	AGC	0V	n/c	n/c	Tx Mute

Connections to the unit are by two 0.1 inch pitch single in line connectors:

- Tx Mute: 0V to inhibit
- Tx Det: DC proportional to dB output power (log detector) Max ~ 4V
- AGC: 0 - 5V Control - from the data sheet turning the power down more than 10dB will limit the output power (I have not tried this, just left the pin o/c)

Local oscillator power required is +10dBm (at 12.24GHz). Around -22dBm Tx drive will give you full output (at max gain). Both receive and transmit RF ports are WR-42 (wg20) waveguide.

Availability

Modules were appearing on Ebay in 2009 mainly from dealers in Israel. Any additional modifications will be posted on the Internet [4].

Converting the Ceragon 7GHz Module for 5.76GHz [5]

Fig 4: The Ceragon 7GHz module that can be modified for 5.760GHz.

These will work after modification on 5760MHz. They consist of a receive LNA, mixer & IF amplifier, plus a transmit filter, amplifier & output monitor. Thus just needing an LO, Tx mixer, and a couple of relays to complete a 5.76GHz transverter.

Measured performance on receive with a 432MHz IF, is as follows:

- Conversion gain: +19dB
- System noise figure: 2.6dB
- Image rejection: (5.328GHz LO) -22dB

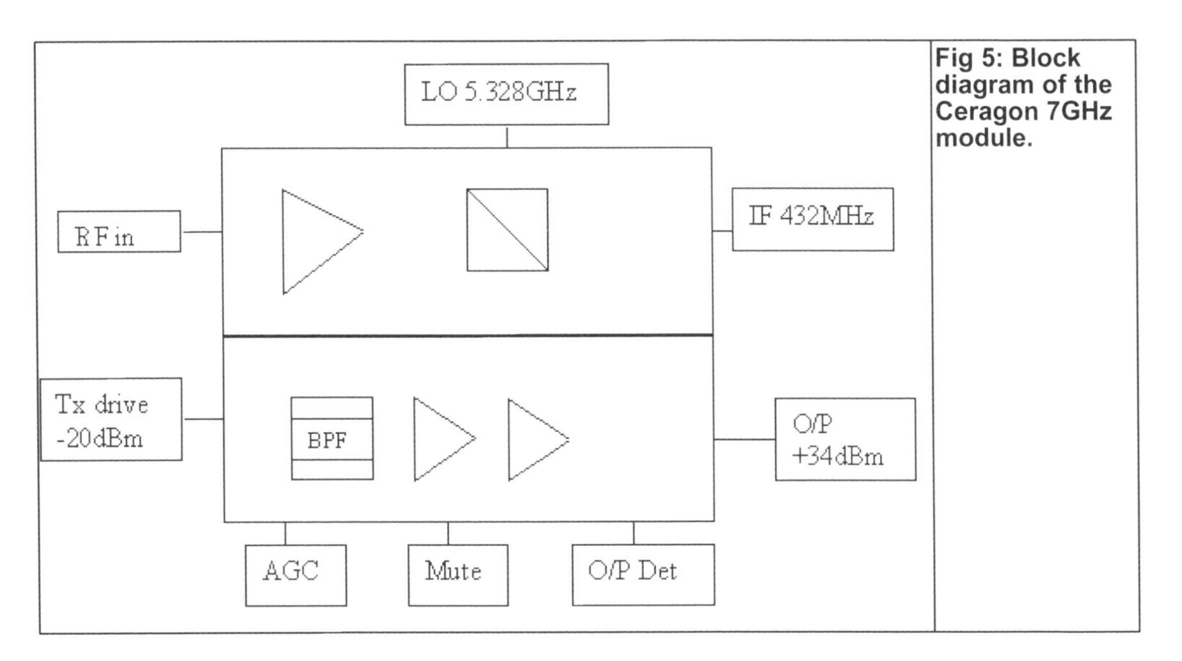

Fig 5: Block diagram of the Ceragon 7GHz module.

Performance on transmit:

- Output: +33.7dBm (2.3W) - 1dB compression
- Saturated output: +34.3dBm (2.7W) saturated
- Gain: +53.5dB

Supply requirements:

	Receive (Tx Inhibited)	Full Output
+5.0V	0mA	50mA
+6.8V	180mA	0mA (switched)
+10.0V	0mA	2.1A
-5.0V	40mA	40mA

On the transmit side

The only essential modification is to place a piece of ceramic on top of the existing inter-digital filter. The unmodified and modified transmit side are shown in Figs 6 and 7. This shifted the entire frequency response down about 1.5GHz, to exactly where needed. Transmit image was 60 – 80dB dependant on the size of the ceramic. With the high transmit gain, it is necessary to either use an input attenuator, or use the AGC control to turn the gain down. This will give 1dB compression of around +32dBm. Adding some small tabs around the PA pushed the power up to +33.7dBm, see Fig 10.

On the receive side

The image filter strips were extended by about 2mm using silver paint. The unmodified and

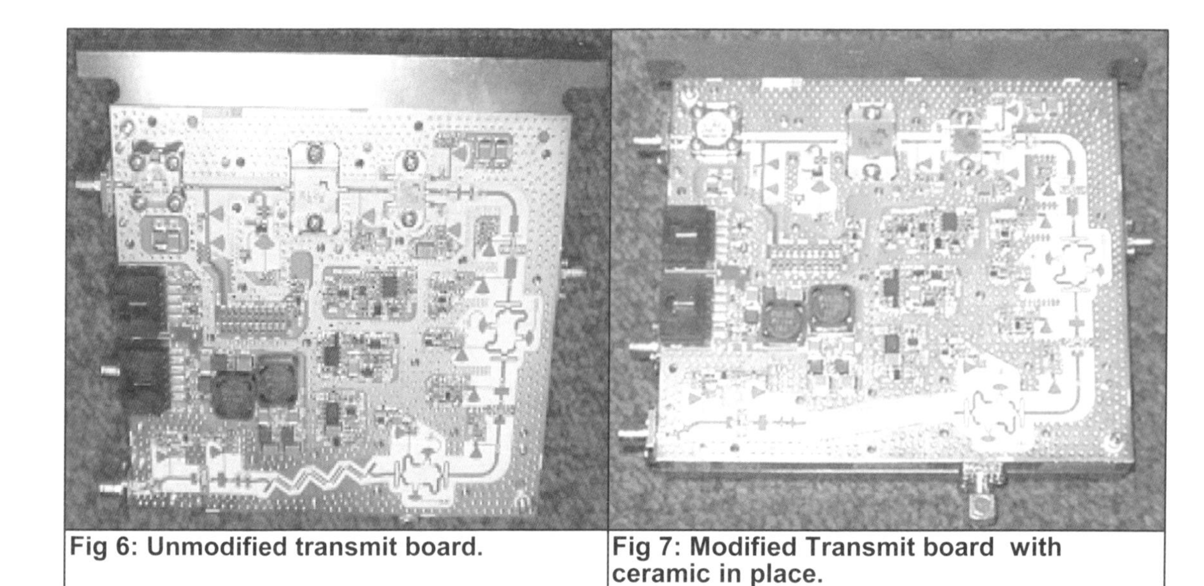

Fig 6: Unmodified transmit board.

Fig 7: Modified Transmit board with ceramic in place.

modified receive side are shown in Figs 8 and 9. This gave 22dB image rejection, but conversion gain was only about 11dB. Starring the RF amplifiers with small pieces of copper, improved the gain by about 7dB

The IF response was about 1 dB down at 432MHz, so the IF output capacitor was changed to 330pF to give a final conversion gain of 19dB.

In my 5.7GHz system I use the Ceragon block with an Alcatel synthesiser switched for either 10 or 5.7GHz, using a 10MHz reference.

No information was available on pin-out or supply voltages, so the following is my derivation.

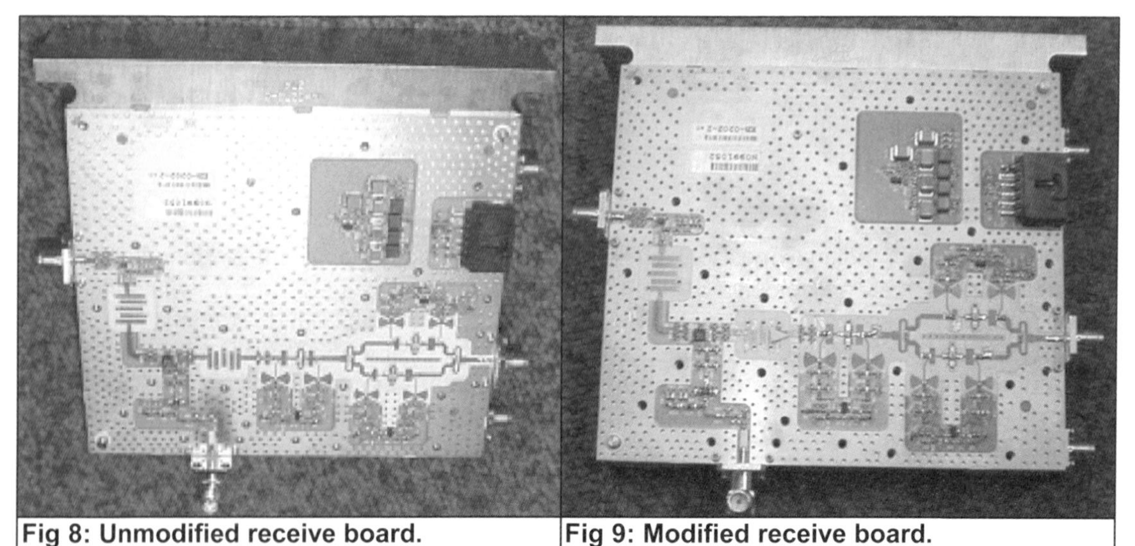

Fig 8: Unmodified receive board.

Fig 9: Modified receive board.

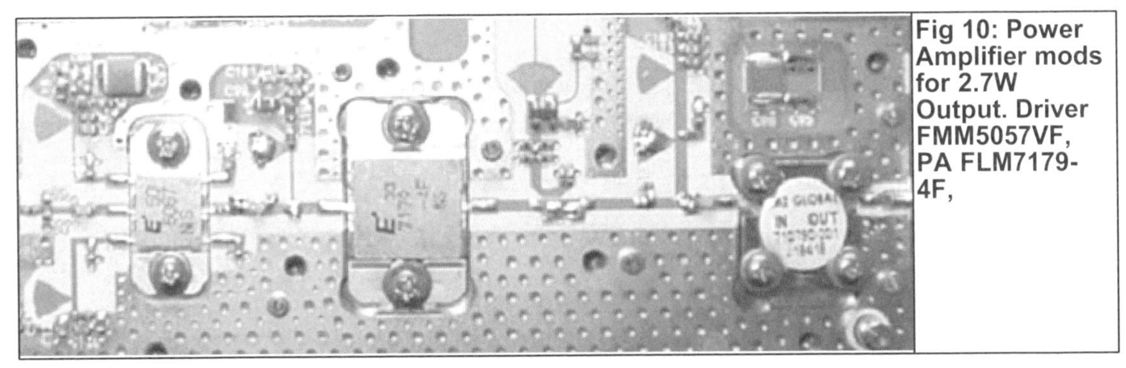

Fig 10: Power Amplifier mods for 2.7W Output. Driver FMM5057VF, PA FLM7179-4F,

From data on the TX driver, a negative supply of -6V may be safer, although I used -5V.

Pin-out on the two connectors as shown in Fig 4, left to right, is as follows:

Rx Connector 1 (6 way)

1	2	3	4	5	6
n/c	+6.8V (IF)	-5V	0V	+6.8V (RF)	n/c

Tx Connector 2 (6 way)

1	2	3	4	5	6
n/c	+10V (driver)	-5V	0V	+5V (amps)	+10V (PA)

Connector 3 (5 way)

7	8	9	10	11
n/c	Tx Det	AGC	0V	Tx Mute

Connections to the unit are by two 0.1 inch pitch single in line connectors. It is quite easy to solder wires straight on if you don't have suitable mating plugs.

In my system, the 6.8V supply is provided by a 7808 regulator and dropper resistor. The two 10V supplies use 3 x 10V 1A low voltage drop regulators. The -5V is provided by a block DC - DC converter and -5V regulator.

External protection is required, to inhibit the positive supplies if the -5V supply fails:

- Tx Mute: 0V to inhibit
- Tx Det: DC proportional to dB output power (log detector) Max ~ 4V
- AGC: 0 - 5V Control I guess (I have not tried this, just left the pin o/c)

Local oscillator power required is +10dBm (5.328GHz). Around -22dBm Tx drive will give you full output (at max gain). The two SMA output connectors allow easy use of a coax relay or the addition of a single stage low noise amplifier or PA.

Availability

Units are currently available on eBay [6] however they may have some faults, particularly Rx FET failure, and damaged SMA connectors.

DXR – 700 TRV conversions to 5.7 GHz. [7]

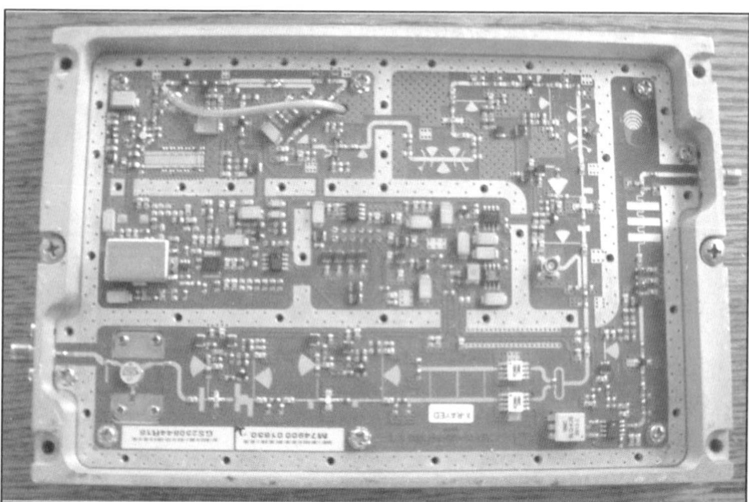

Fig 11: The DXR - 700 unit suitable for conversion to the 5.76GHz amateur band.

The unit shown in Fig 11 was made by DMC (7.1 GHz) and can be converted to the 5.76GHz amateur band using a 144 MHz IF. The intention of this article is not to be specific but to explain how easy they are to convert. This may assist entry level / beginners workshops.

These units have been available for a number of years from the Wellington VHF group here in New Zealand [8].

Rather than use the original PLL I have used crystal oscillator injection at 1123.2MHz and feed this into the onboard multiplier chain on both the TX and RX boards.

A 93.6MHz crystal is used with an EME 65 kit from Mini-kits Australia [9]. The output from 561.6MHz is feed into a Waikato VHF doubler board using an ERA 3 as a multiplier. Output at 1123.2 is then feed to both TX and RX boards of the DXR – 700, see Fig 12.

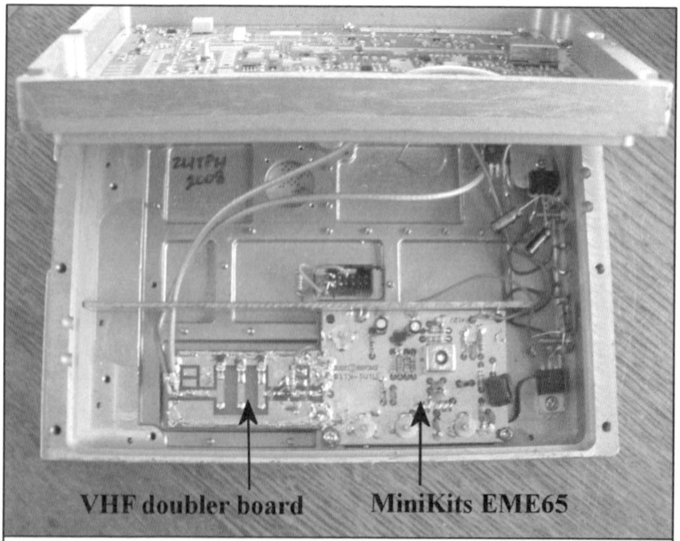

Fig 12: MIniKits EME 65 multiplier board and VHF doubler board feeding the TX and RX boards.

Voltage connections to the DXR – 700 are via pins internal in the casing where the oscillator chain is housed. I use three 9 volt regulators and these are each used to feed the TX, RX and oscillator boards. Just three pin are used for DC connections on each board.

Because the on board filters in the multiplier chain are no doubt tuned around the original 7GHz frequency they do seem to somewhat unforgiving at 5616MHz. Only one stub

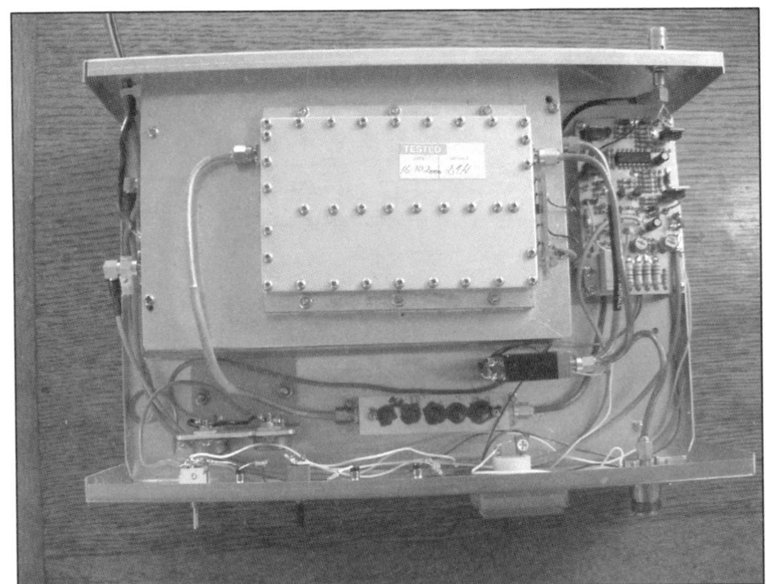

Fig 13: The completed 5.76GHz transverter.

requires a tab and that is the filter before the final GaAs-FET in the multiplier chain.

Because the 1123.2MHz oscillator injection will also multiply by 6 to 6739.2MHz (which we do not want) a 5.7GHz narrow band pass filter is required on TX. Here in New Zealand we use the 7.1GHz filters returned to 5.7GHz by lengthening the posts with a blob of solder.

Effectively at this point we have an easy 10 ~ 20mW TX output on 5.7GHz and RX N/F of about 6dB. For connections to your 144MHz IF (ICOM 202 of FT 817) I use a sequencer from Minkits which gives a host of functions such as RX /TX drive levels and DC switching etc. As can be seen in Fig 13 I use one of the 7.1GHz power amplifiers and this gives a TX output of 5W.

Improvements would be a good RX preamp on receive and post 5.7GHz filter. But without these we here in New Zealand have managed 350km contacts using only 60cm dishes. The conversion of one of these units is relatively easy for the newcomer / entry level to microwave construction with assistance from others in the UK microwave group [1].

Eyal Gal 11GHz Transceiver [10]

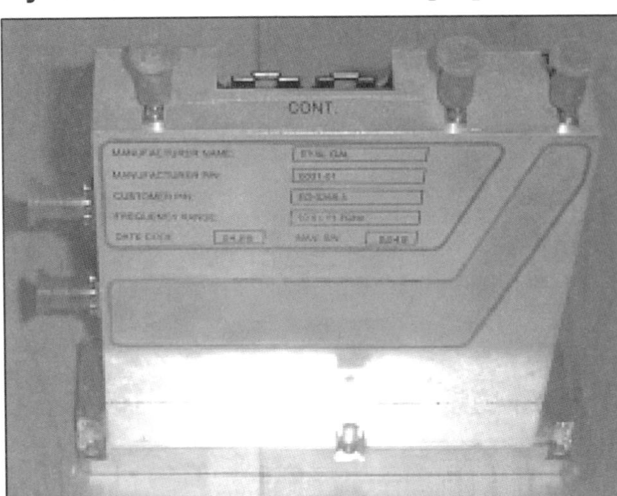

Fig 14: The Eyal Gal 11GHz transceiver that will work on the 10GHz amateur band.

These units with the part number 6031-01 will work unmodified at 10.368MHz see Fig 14. They consist of a receive LNA and mixer, plus a transmit amplifier and output monitor. Thus just needing an LO, Tx mixer and filter and a couple of relays to make a neat 10GHz transverter.

Measured performance on receive with a 432MHz IF, is as follows:

- Conversion gain: +22dB
- System noise figure: 3.9dB
- Image rejection: (9.94GHz LO) -24.5dB

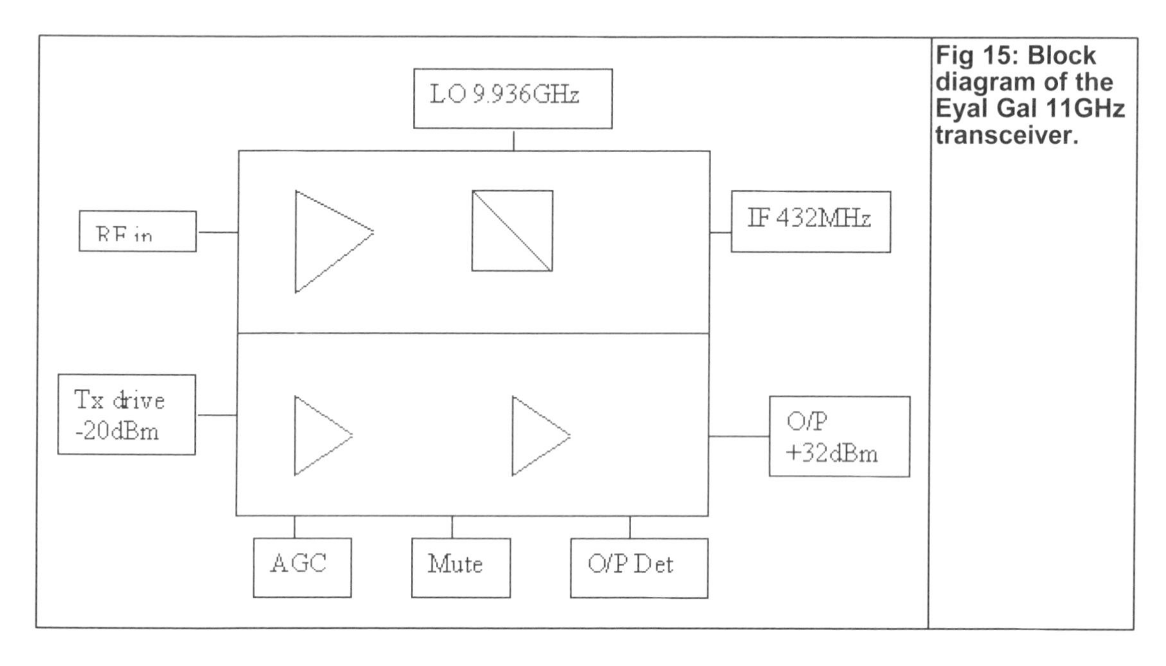

Fig 15: Block diagram of the Eyal Gal 11GHz transceiver.

Performance on transmit:

- Output: +32dBm output -1dB compression
- Gain: +53dB gain
- Saturated power output on transmit is >2W.

Supply requirements:

	Receive (Tx Inhibited)	Full Output
+8.0V	380mA	720mA
+12V	50mA	1.93A
-12V	105mA	105mA

The IF response is flat from 75 -1700MHz, but a 144MHz IF would only give you 2dB image rejection, so is not practical. With the high transmit gain, it is necessary to either use an input attenuator, or use the AGC control to turn the gain down.

In my 10GHz system I use the Eyal Gal block with an Alcatel synthesiser and 10MHz reference. The block dissipates a fair amount of heat, so don't remove the aluminium slab base plate. As you can see in Fig 16 showing my complete transverter, I have done just that. But, I have replaced it with an L-bracket to the base plate, which works just as well.

A specification sheet was available on the Internet [11] but no connection details, so I had to work those out.

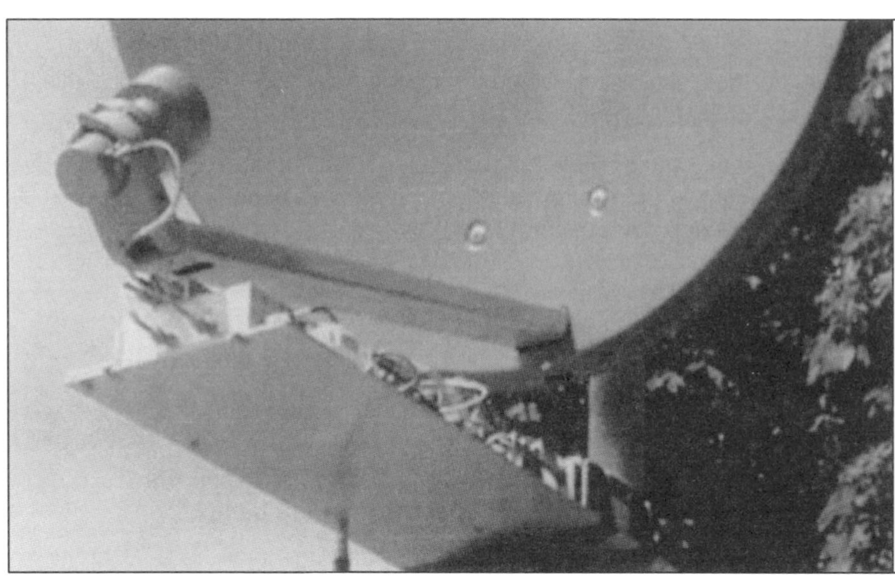

Fig 16: Complete 10GHz transverter in operation.

The output power measured was much greater than that indicated by the specification sheet (harmonics are quoted with a +26dBm power, and IP3 as >+38.5dBm). I can only put that down to the fact that I am using it at the bottom end of its frequency range.

Pin-out on the two connectors as pictured above, left to right, is as follows:

Connector 1 (6 way)

1	2	3	4	5	6
n/c	+8V	-12V	0V	n/c	+12V

Connector 2 (5 way)

7	8	9	10	11
n/c	Tx Det	AGC	0V	Tx Mute

Connections to the unit are by two 0.1 inch pitch single in line connectors. It is quite easy to solder wires straight on if you don't have suitable mating plugs.

In my system the 8.0V supply is provided by a 7808 regulator. The 12V supply is a direct battery feed. Both are internally regulated, so the exact voltage is not critical. The -12V is provided by a block DC - DC converter.

I have not had the courage to remove the -12V supply to see if the unit is internally protected, but equally I have not blown one up yet.

- Tx Mute: 0V to inhibit
- Tx Det: DC proportional to dB output power (log detector) Max ~ 4.3V

- AGC: 0 - 5V Control - from the data sheet turning the power down more than 10dB will limit he output power (I have not tried this, just left the pin o/c)

Local oscillator power required is +3 to +6dBm (9.936GHz). Around -21dBm Tx drive will give you full output (at max gain). The two SMA output connectors allow easy use of a coax relay, or the addition of a single stage low noise amp and/or PA.

Availability

Always when writing about something, that gets published some time later, there is a danger that supply has disappeared. Try searching eBay initially, or surplus suppliers in Israel.

Using the RX converter of White Boxes [12]

Fig 17: The White Box converter.

Fig 18: Base of the White Box converter.

Using the RX converter on the 24.048MHz amateur band.

Many of us can now buy the 23GHz microwave links called "boite blanche" (white box) at low cost.

I have been checking the RX converter. Several types are available, all very similar:

- GBX124 : 23GHz RX converter with LSB mixer (LO frequency lower than RX frequency).
- GBX125 : 23GHz RX converter with USB mixer (LO frequency higher than RX frequency).
- GBY111 like 125

More information (pictures, schematics etc.) can be found on the Internet [13].

If your converter is an LSB type there is no problem, you can use it without modification (generally there is an INF sticker on the synthesiser box).

My White Box was a GBY111 (SUP sticker) so without modification the LO had to be on 12240MHz (harmonic mixer and 432 IF) If I had used the LSB LO (23616MHz) the 24048MHz would be rejected (image reject mixer).

Fig 19: The electronics inside the White Box.

RX converter modification to use the LSB LO on an original USB converter

The mixer used in the White Box converter is an image reject type. Two mixing products can be found on the 3dB output coupler. On one of the outputs there is the LSB product, on the other the USB product. So all that is required is to use the other output.

To open the converter box insert a small screwdriver tip between the two covers, both covers are only silver pasted.

The electronics can be seen once the box is open see Fig 19. On the left hand side there is the WR42 port and its microstrip adapter. The hybrid module is the 23GHz low noise preamplifier.

Fig 20 shows the circuit diagram. There is a 3dB coupler on the preamplifier output driving two harmonic mixers (the LO is 12GHz). Both mixers are summed in the 3db quadrature output coupler. One of the two outputs is terminated by R1. The other output is connected to the IF amplifier via R2 (in my case it was a 22Ω, not 18Ω as shown on the circuit diagram). The remaining components are for polarisation and LO.

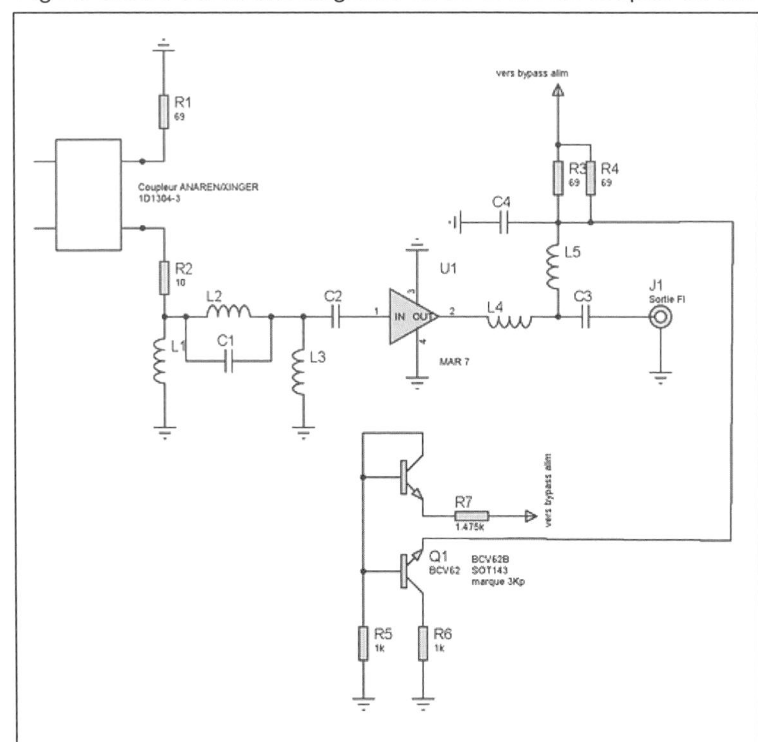

Fig 20: The circuit diagram of the White Box converter.

If your converter is not the right version the two Anaren coupler outputs must be reversed. The output originaly connected to the IF amplifier will go to ground via a 68Ω resistor (SMD resistor in my case) and the other one will go to the IF amp via 22Ω. This is shown in Fig 21:

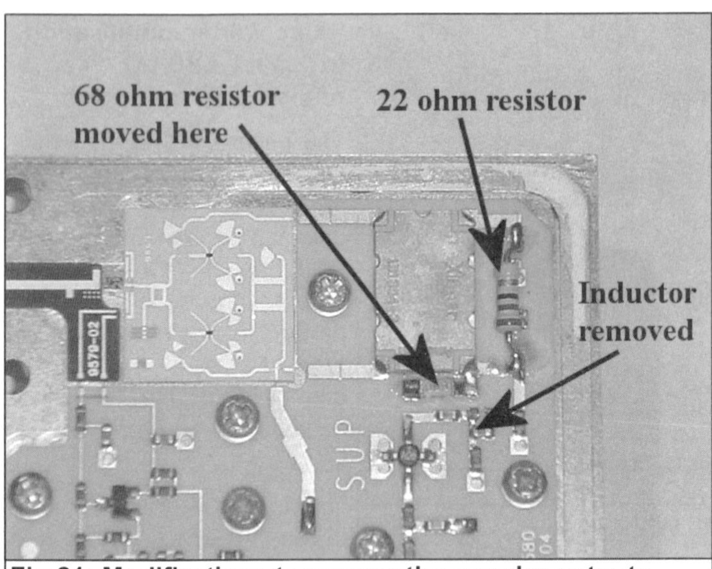

Fig 21: Modifications to reverse the coupler outputs.

- Carefully unsolder the two resistors then solder a 68Ω SMD resistor between the 3db coupler output and the closest via to ground. Then solder the 22Ω resistor on the other side to the IF amplifier input.

IF passband

Originally the White Box IF was around 600/800MHz. In my case I wanted to use 432MHz for an IF. The question was: was this possible?

Two possible problems arise:

- The 3db Anaren coupler was not designed to be used at 432MHz. What will the image rejection be if the same coupler is used with a 432MHz IF?
- Is the IF amp useable at 432MHz?

First of all I ran a network analyser measurement of the IF amplifier, this is shown in Fig 22.

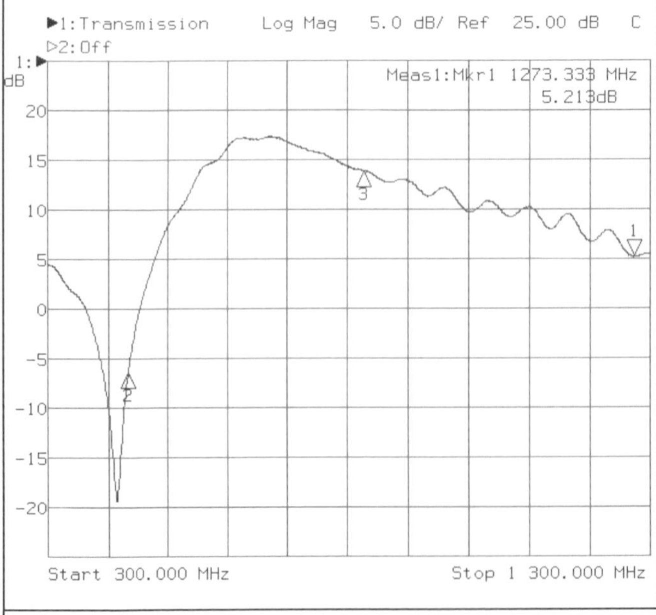

Fig 22: Network analyser plot of the IF amplifier.

From this we can see that the IF amp gain is not OK at 432MHz, the gain is -7db. I decided to remove the inductor L2 inductor because it creates a rejection pole with C1. This gave 17db gain at 432MHz that is much better as shown in Fig 23.

Image rejection

A good idea would be to replace the 800MHz Anaren coupler with the same model but designed for 432MHz. Unfortunately I only had one converter available. Removing the original coupler is a risky operation with no spare converter so I decided not to try. A question remained: what was the image rejection? Figs 24 and 25 show that we can state the image rejection is 9db, not extraordinary, but acceptable.

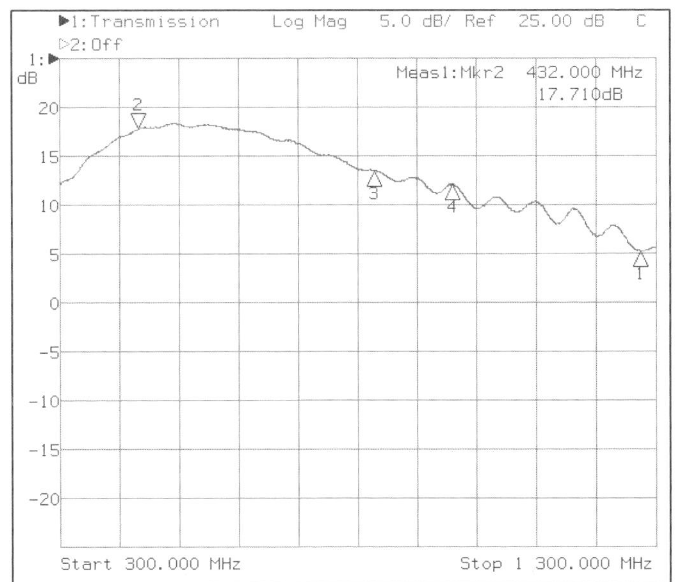

Fig 23: Network analyser plot of the IF amplifier after modification.

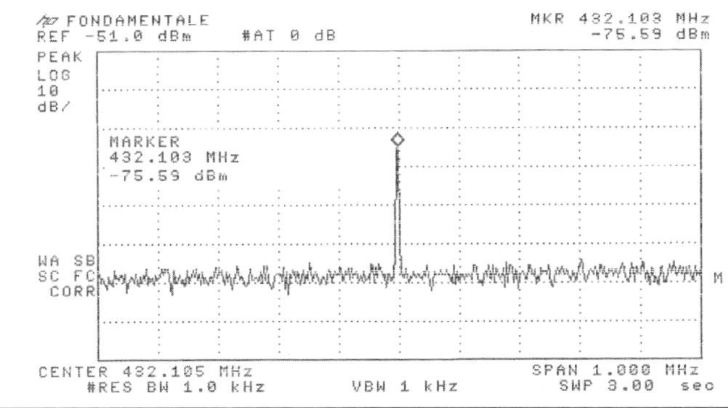

Fig 24: With 24048MHz at the input the 432MHz output is -75.6dBm.

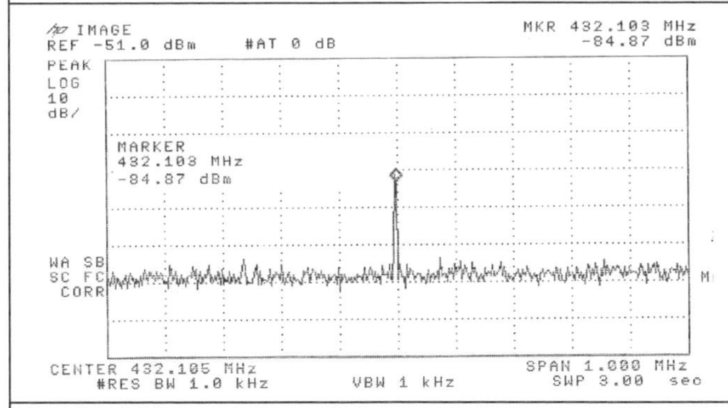

Fig 25: With the image frequency at the input (same level), the 432MHz output is -84dBm.

Noise figure

I ran some cold sky/ground measurements to do some NF evaluations doing an average of several measurements. The converter noise temperature measured is about 270°K. But there is some image noise contribution to take into account. What we measure here is the DSB noise figure. What we need is the SSB noise figure so we have to correct our measurement. Usually we don't need to do this because the image rejection on the systems we run CS/GND have enough image rejection to neglect it. If we do the correction the SSB noise temperature is about 300°K giving about 3db NF.

Conclusion

The White Box converter gives good performance. For a top-notch 24GHz RX (1.5db NF) a preamplifier should be used with enough gain to be able to add an OE9PMJ filter between the converter and the preamplifier to reduce the image frequency. The references [14] and [15] were used for measurements. The following colleagues are acknowledged for their help with this article: F1VL, F1BOH, F1CHF, F6BVA, F9HX and G4ALY.

References

[1] UK microwave group, http://www.microwavers.org/

[2] Eyal Gal 21.2 – 23.6GHz Transceiver, Roger Ray G8CUB, Scatterpoint February 2009

[3] Specification for Eyal Gal 21.2 – 23.6GHz transceiver is available on: http://www.eyal-emi.com/siteFiles/1/32/1081.asp

[4] Further modifications for the Eyal Gal 21.2 – 23.6GHz transceiver are available on: www.rfdesign.co.uk/microwave, Roger G8CUB email: littlemallards@hotmail.com

[5] Converting the Ceragon 7GHz Module for 5.7GHz, Roger Ray, G8CUB, Scatterpoint November/December 2008

[6] Ceragon modules were available from eBay seller art-in-part in 2008.

[7] DXR – 7—TRV conversion to 5.7GHz, Stephen Hayman, ZL1TPH, Scatterpoint September 2008

[8] Wellington VHF group - http://www.vhf.org.nz/pubs/TradingTable/mwave6.pdf

[9] Mini Kits - http://www.minikits.com.au/

[10] Using the Eyal Gal 11GHz Transceiver on 10GHz, Roger Ray, G8CUB, Scatterpoint July/August 2008

[11] A specification sheet for the Eyal Gal 11GHz transceiver was available from - http://www.eyal-emi.com/siteFiles/1/32/1081.asp

[12] 24GHz Whiteboxes, Dom Dehays, F6DRO, Scatterpoint March 2008

[13] Information on White boxes can be found here - http://f1chf.free.fr/boite%20blanche/forum.htm

[14] Principe de la mesure du facteur Y et calcul du NF par F5CAU.

[15] Image reject mixers by Aksel Kiis in Applied Microwave, Winter 91/92

Chapter 6

Converters

In this chapter :

- S Band (13cm) Receive Converter
- 144MHz to 28MHz receive converter

onverting signals from one band to another is a common technique to use an existing receiver or transceiver to listen to a different amateur band. The converters in this chapter show two different reasons for this conversion process.

S band (13 cm) Receive Converter

For operation on AO-51 and other amateur satellites it is convenient to have a receive converter that converts signals in the 2400MHz band to UHF. This satellite and others typically operate in V/U, L/U or L/S and occasionally V/S that is VHF up link and UHF down link, L band (23cm) up link, UHF down link, or L band up link, S band down link, and VHF up link with S band down link. (Note that the terms S band and L band are somewhat archaic. The correct modern designations are D and E bands.) It is useful for the latter modes to have an S band to UHF receive converter to minimise the amount of re-tuning and setup changes on the transceiver. A quick survey of the Internet reveals few or no commercial suppliers of such transverters, as kits or prebuilt, for the amateur market; hence the motivation for the present project.

Any receive converter consists of two primary sections a synthesiser or other stable local oscillator and a converter section consisting of a mixer, one or more amplifiers and one or more filters. The synthesiser used here is a fairly standard microprocessor controlled unit tuneable, under software control, to any frequency between 1 and 2GHz. The microprocessor is a "large" unit with many spare pins; this was an intentional choice to allow expansion to sequencing or multi-band operation. In addition with simple software changes, the serial interface on the microprocessor can be enabled to allow remote software controlled tuning, sequencing or other functions. In the present design however, the software is very simple and sets one of two fixed frequencies depending on the state of one of the spare I/O pins. The two pre-programmed frequencies are 1966MHz and 1152MHz – suitable for S-U and L-V conversion respectively. The synthesizer employs a PIC 16F57 microprocessor that can be easily programmed in circuit via an inexpensive PIC-KIT programmer, or a variety of commercial or home brewed programming interfaces. The schematic specifies an Analog Devices ADF4112 synthesiser however there is a pin compatible National Semiconductor device available as well. This can be used with only minor software changes. The VCO is a ROS-2150 from mini-circuits. All of the components, except the VCO and RF buffer amplifier are available through mail order from Digikey [1] – this was actually one of the design goals: to use easily available components. The VCO and amplifier can also be obtained mail order directly from the manufacturer in single unit quantities. The Gerber (RS-274) files needed for having the circuit board professionally fabricated are available from my website [2]. The synthesiser is shown in the Figs 1 - 5. The initial prototype had some minor errors that required a few tracks to be cut and wires to be

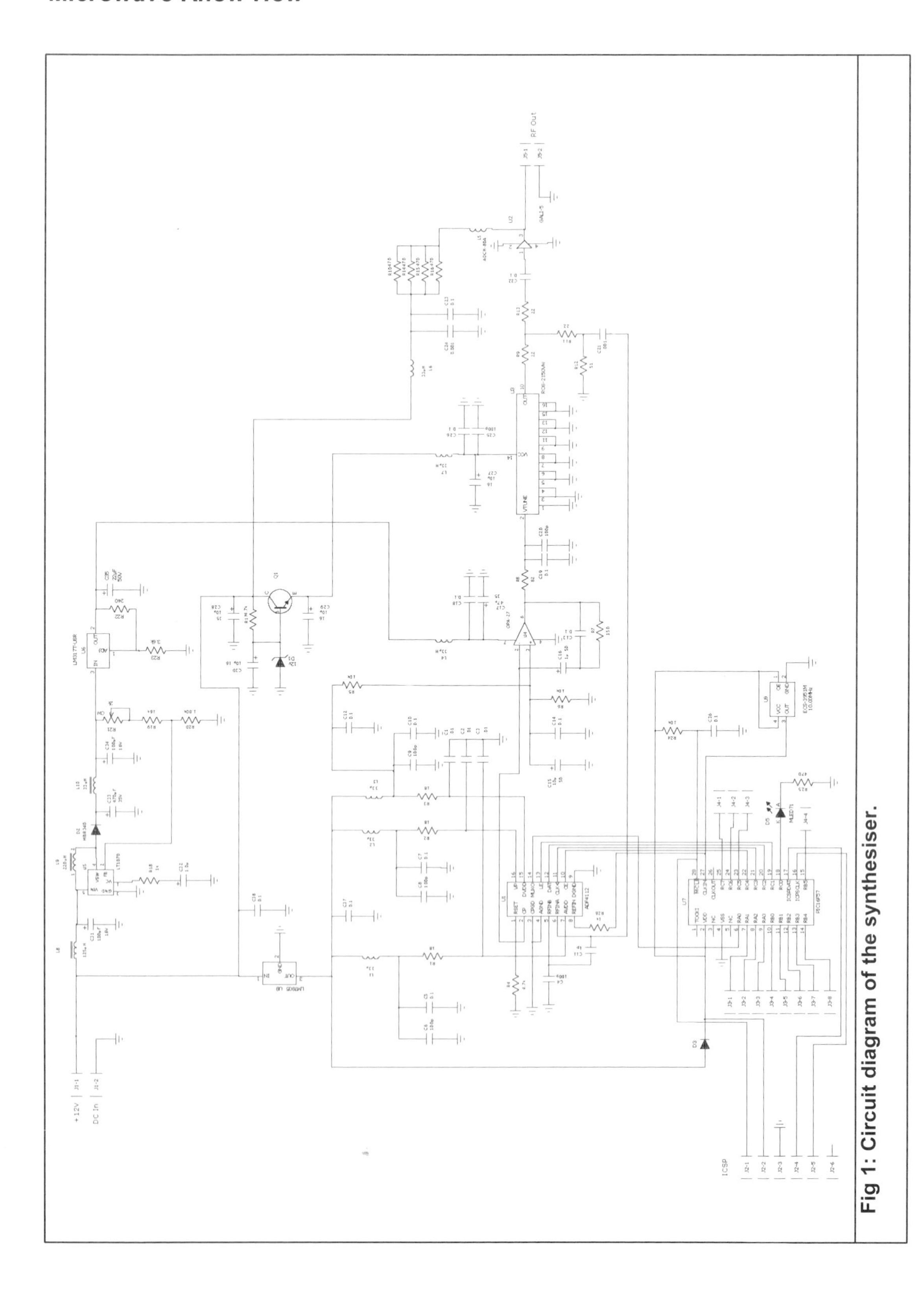

Fig 1: Circuit diagram of the synthesiser.

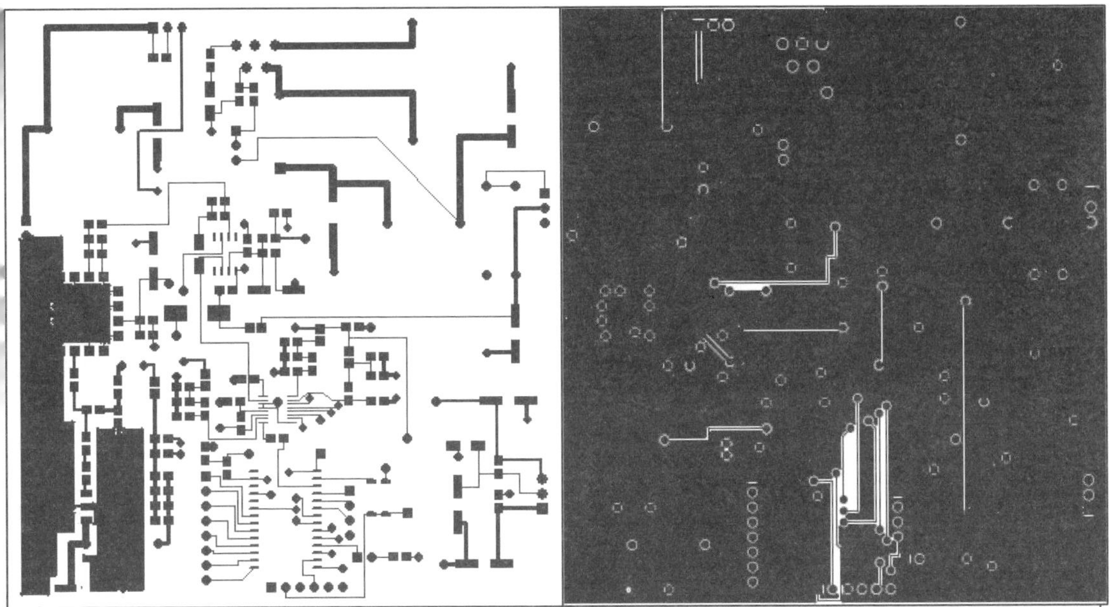

Fig 2: PCB layout for the top of the synthesiser board.

Fig 3: PCB layout for the bottom of the synthesiser board.

applied. These have been corrected in the circuit diagram, and PCB layouts on the web site [2] and included in this article.

Now that we have a synthesiser we can proceed to the converter section. I happened to find a very low noise preamplifier for the converter on eBay for a low price so I used it instead of incorporating a preamplifier into the converter itself. The converter can be used successfully

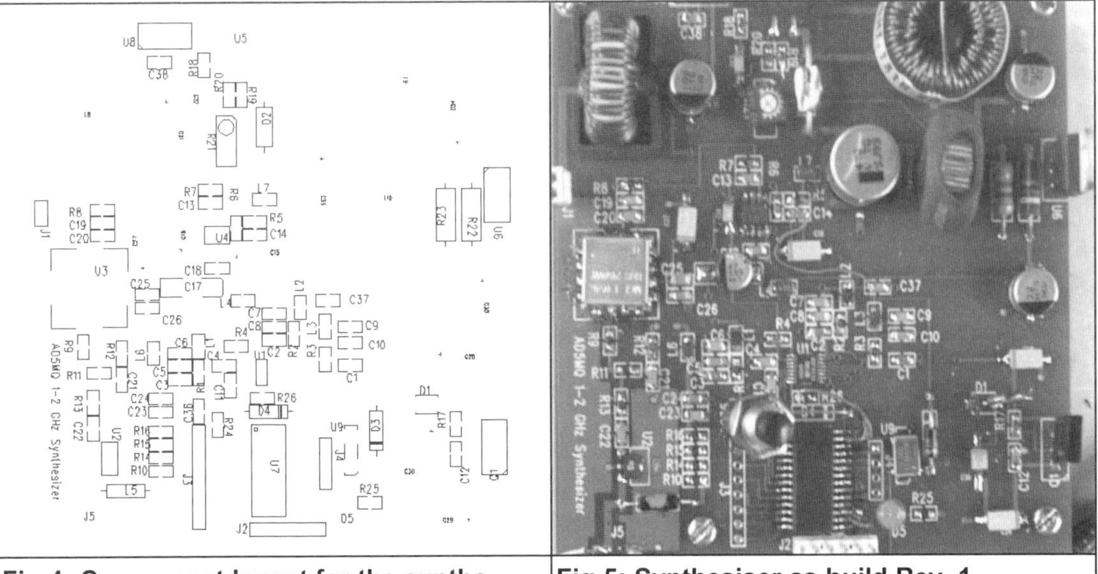

Fig 4: Component layout for the synthesiser board.

Fig 5: Synthesiser as build Rev. 1.

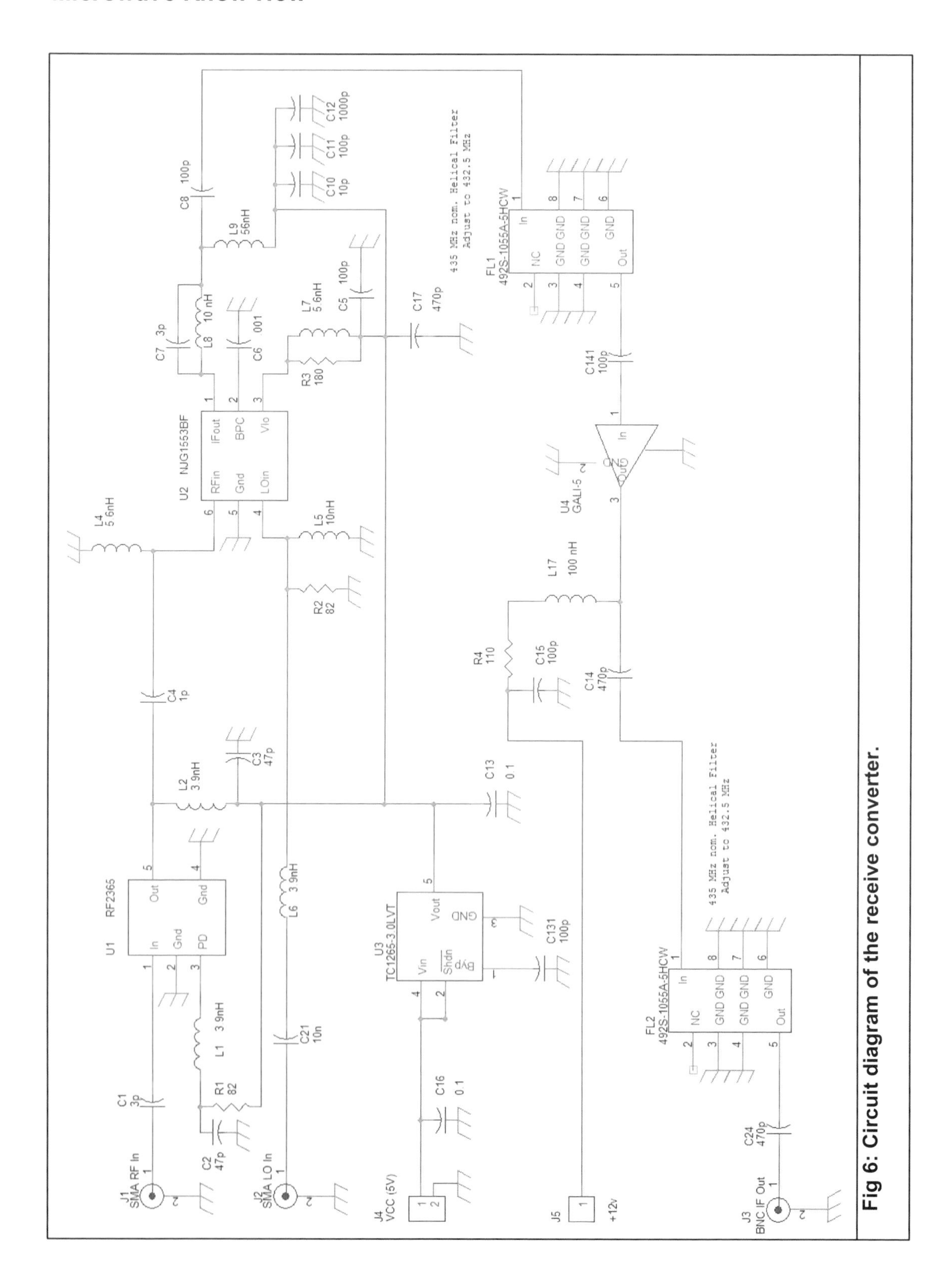

Fig 6: Circuit diagram of the receive converter.

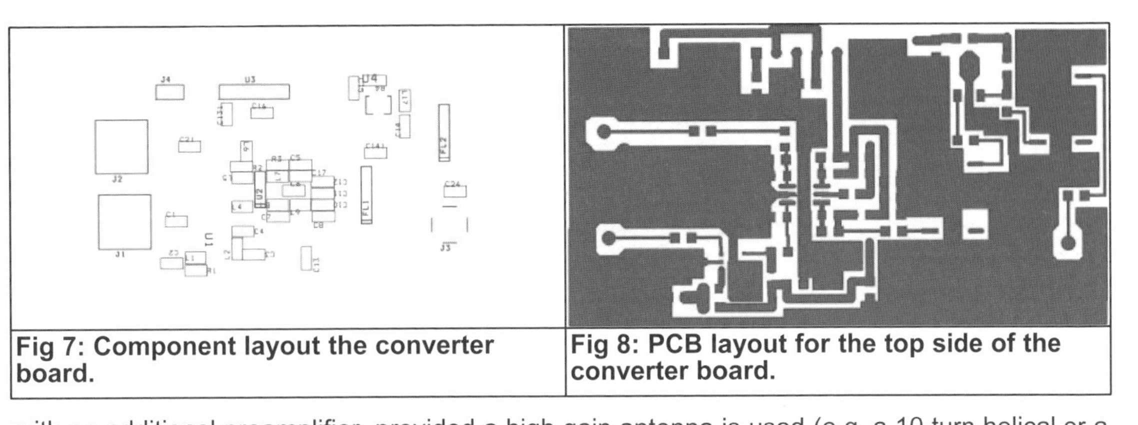

Fig 7: Component layout the converter board.	**Fig 8: PCB layout for the top side of the converter board.**

with no additional preamplifier, provided a high gain antenna is used (e.g. a 10 turn helical or a converted C band TVRO dish). The measured MDS (Minimum Discernable Signal) with no preamplifier is -115dBm. This is somewhat low, but usable – for comparison a typical UHF receiver has a MDS of -121dBm. The converter uses somewhat "old" (more than 3 years) cell phone parts including the RF2365 (RF Micro Devices GaAs LNA) and a NJG1553 (New Japan Radio Corp.) GaAs PCS band mixer which works quite well at 2400MHz. Because these parts are so "old" they may be harder to find now, though I was able to purchase several of each very inexpensively on eBay. Several devices are available from Digikey that are suitable replacements. These are even available in the same packages however they do have different pin-outs. In my converter I opted to use .010" thick (.254mm) Rogers RT-Duroid TM. This is a very low loss material designed specifically for use in microwave applications. It has a very low loss tangent and a very small temperature coefficient of capacitance, both highly desirable in microwave circuits that will be used outdoors. Small quantities can be obtained from Surplus Sales in the U.S. [3]. Because of the relatively low frequencies involved, and the small dimensions of the highest frequency portions of the circuit it is acceptable to use standard 0.031" (0.787 mm) FR4 material instead. If that substation is made, the strip lines at the inputs, outputs and interconnects carrying RF will have to be 0.050" (1.27mm) wide instead of the smaller 0.01" (0.254mm) dimension on the RT-Duroid. The standard 0.062" (1.575mm) FR4 used in conventional digital circuits and indeed in the synthesiser is not suitable for the converter. The 50Ω strip lines are 0.100" (2.54mm) wide on that material and this is MUCH larger than the dimension of the active components, making large, geometrically tricky transitions necessary and introducing significant impedance discontinuities at each active devices input and output. Although similar problems occur on the 0.031" (0.787mm) material they are more tractable and the discontinuities can be made acceptably small, if smooth adiabatic transitions are used. The circuit diagram, PCB layout, component layout and a photo are shown in Figs 6 - 9.

As can be seen in the photo in the final unit I left out the output filter. Also after attaching an external preamplifier I found that the additional gain provided by the IF amplifier was unnecessary and contributing to saturation in the UHF final receiver so I bypassed it. In applications that do not use an additional preamplifier the IF amplifier stage will be needed. The series resistor will need to be adjusted for the particular amplifier selected. Amplifier selection is non-critical since almost any MMIC amplifier will provide good gain and reasonably low noise at 435MHz.

Finally we need to assemble everything into an enclosure, and add a few discrete components

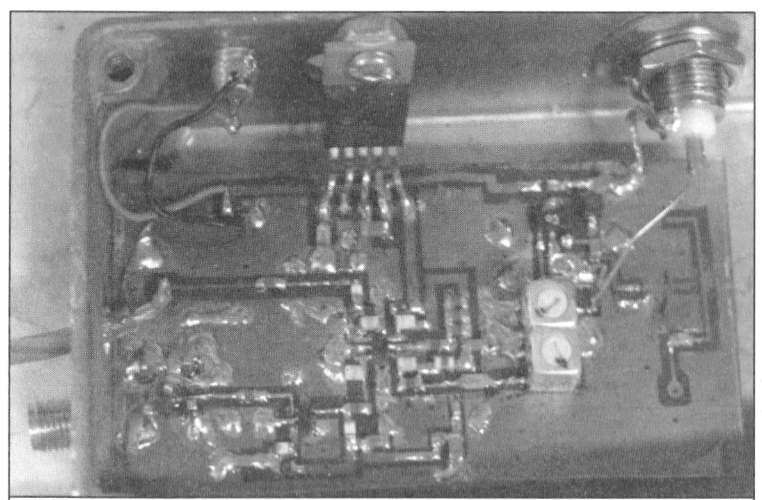

Fig 9: Assembled receive converter.

for DC power take off from the UHF output, or a separate supply, and a filter for the input. Although a preamplifier is optional a good narrow filter is vital to a reasonably low noise figure that will be needed to work spacecraft. Since noise is directly proportional to bandwidth we'd like the lowest practical bandwidth that we can achieve with simple components. A cavity filter is suitably small, easy to construct and has a reasonably narrow bandwidth. If you do not have access to a network analyser, or even a signal generator and power meter for the 13cm band you can still build this filter and have confidence that it will work as long as the dimensions are followed carefully. The filter is constructed from materials that can be found at any hardware store; a few copper plumbing parts, some copper sheet and 2 SMA connectors (of course you can't usually get these at the hardware store!). Figs 10 – 12 show a dimensioned drawing and a couple of photos of the assembled filter.

The filter is made mostly with standard copper plumbing fittings. The cavity itself is made from a standard ¾" pipe cap soldered directly to a ¾" coupler. Make the cavity first – it will require use of a small torch to get enough heat to solder the cap to the coupler. These will have to be clamped when starting to solder with a metal C - clamp. Because, in plumbing nomenclature, pipe size is based on inside diameter and because these fittings fit over the outside of the pipe the implied dimensions are misleading. The actual inside diameter of the cavity is 0.875" and

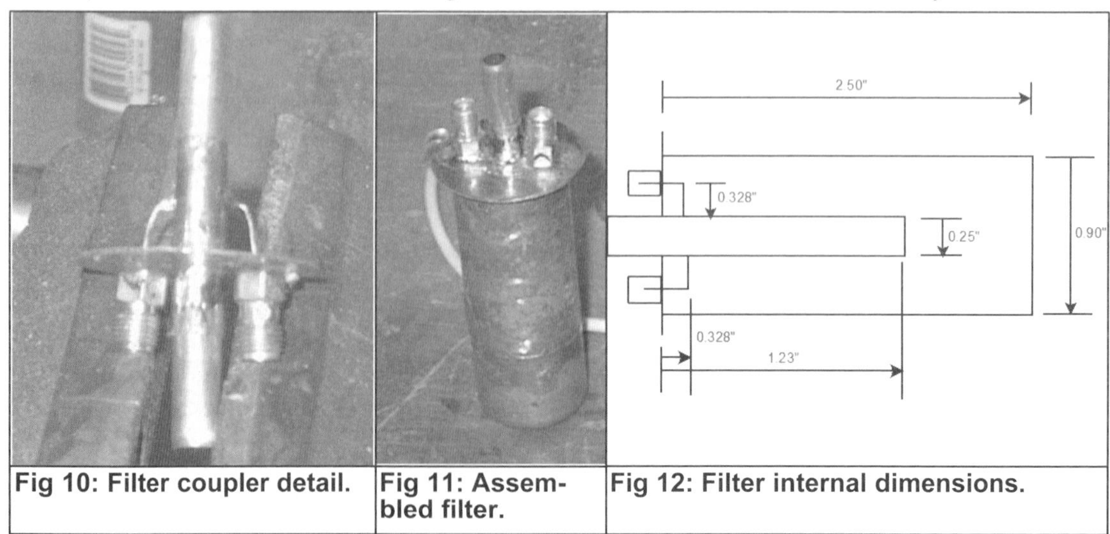

Fig 10: Filter coupler detail. Fig 11: Assembled filter. Fig 12: Filter internal dimensions.

the outside diameter is slightly more than 1". The connector plate is a disk of copper sheet (thickness non-critical) about 1.25" in diameter with a 0.240" hold drilled in the centre and the smallest holes which will pass the SMA centre pin for your connectors – nominally 0.453" from the centre. The coupling loops are nominally 0.328" square, this is not very critical; 0.050" variation makes little difference in the response. The tuning stub is formed from standard ¼" copper tubing (tubing, unlike pipe, is measured by outside diameter so it is 0.25"). Wrap the tubing in a piece of copper sheet approximately 1" x 1.25" – long way around - as tightly as possible and make sure the sheet overlaps itself, then solder the sheet to itself, being careful that none of the solder flows onto the tubing – the goal is to have a tight fitting collar around the tubing which allows the tubing to be moved within it. Insert the collar – with tubing still inside into the hole in the copper disk and solder in place. The solder needs to provide a firm mechanical connection but it is not critical that it flow all the way around the collar. Solder the SMA connectors to the disk. Fabricate the coupling loops from small diameter solid wire and solder in place. Adjust the depth of the tubing to 1.230" from the bottom of the disk and finally tack solder the disk to the cavity, centring the tuning stub as well as possible. If you have access to

Table 1: Parts list for the S band converter.

Part	Value	Package
R1, R2, R3	18Ω	0805
R4, R17	4k7Ω	0805
R5, R6, R24	10kΩ	0805
R7	150Ω	0805
R8	82Ω	0805
R9, R11, R13	22Ω	0805
R10, R14, R15, R16, R25	470Ω	0805
R12	51Ω	0805
R18, R20, R26	1kΩ	0805
R19	16kΩ	0805
R22	240Ω	½W
R23	3k6Ω	½W
R27	100kΩ	0805
R21	1kΩ	Pot
C1,C2, C3	0.01µF	0805
C4, C6, C8, C9, C20, C25	100pF	0805
C5, C7, C10, C12, C13, C18, C19, C22, C23, C26, C36, C37, C38	0.1µF	0805
C11	1pF	0805
C21, C24	0.001µF	0805
C16, C32	1µF/50V 3.3 x 3.3mm cap	
C15, C27, C28, C29, C30	10µF/50V 5.3 x 5.3mm cap	
C17	47µF/50V 5.3 x 5.3mm cap	
C34, C35	22µF/50V 6.6 x 6.6mm cap	
C31	100µF/35V 6.6 x 6.6mm cap	
C33	470Fµ/35V 6.6 x 6.6mm cap	

Part	Value	Package
L1, L2, L3, L4 L6. L7	33µH	0805
L5	82µH	axial lead
L8	125µH	radial lead torroid
L9	220µH	radial lead torroid
L10	33µH	radial lead torroid
D1	13V zener	DO-214AA
D2	3A	Schottky 40V 3A
D3	IN914	Schottky 100mA
D5	LED	
U1	ADF4112BRUZ	
U2	GALI-51+	
U3	Mini Circuits ROS-2159VW+	
U4	OPA-27	
U5	LT1070	
U6	LM317	
U7	PIC16F57	
U9	XC296CT-ND	
Q1	2SC1847	

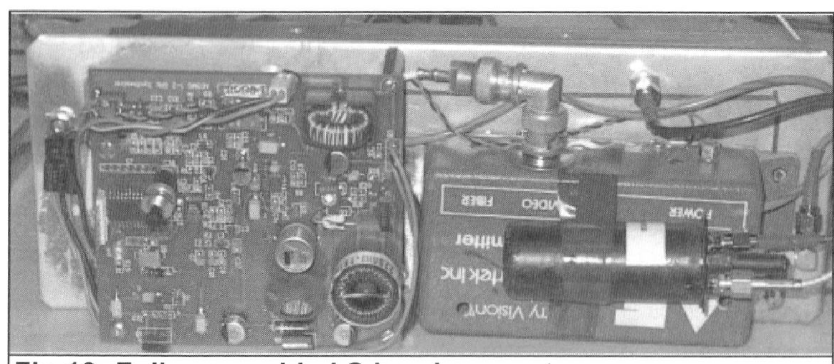

Fig 13: Fully assembled S band converter.

a network analyser, or alternatively a signal generator and power meter, move the tubing in and out slightly to minimise the loss at 2401MHz (for AO-51 a different frequency may be needed for other satellites). The tuning is not terribly critical – the loss is about 6dB 50MHz from centre. This is enough to provide useful rejection for nearby signals such as WiFi bridges without resulting in extremely critical tuning. It is, of course, possible to obtain much narrower responses from more filter stages – multi-cavity or inter-digital filters work well here, but at that point advanced test equipment such as a network analyser is essential. Such narrow filtering would be essential for EME or other extremely weak signal work, but is not needed for most amateur satellites. Fig 13 shows a photo of the final, fully assembled receive converter (with the cover off).

As can be seen in the photo, the 5V regulator on the synthesiser board is mounted to the base plate. This was necessary because the front end amplifier and mixer are powered from it, this raises the current consumption quite a lot on the 5V line necessitating heat sinking the regulator. The front end was assembled into a separate shield enclosure – the visible box is a salvaged case from an old fibre optic video link.

144MHz to 28MHz receive converter [4]

Many of the current Software Defined Radios (SDR) are limited to a maximum tuned frequency of 30MHz. One way to use these SDRs on the VHF, UHF or microwave bands is to use a frequency downconverter, bringing the upper frequencies within the SDR's range. In principle any output frequency up to 30MHz could be chosen for this 'intermediate frequency' (IF), but 28 - 30MHz is a common choice.

An older design of converter, such as the Microwave Modules MMC144, could be pressed into service but these tend to suffer in the presence of strong signals due to their high gain. They also tend to exhibit poor frequency stability and frequency offset limitations, sometimes being unable to net the local oscillator onto frequency. What is required is a low gain converter, with good frequency and gain stability, together with low noise sidebands. A very low noise figure is not required in this application as the converter is usually preceded by a relatively high gain, low noise transverter (for higher bands) or a masthead preamplifier for 2m. I decided to design and build a new 144MHz to 28MHz receive converter that would meet these requirements. The result is the DDK2010 described in this article.

Extra features

In order to make the converter even more useful I decided to add the facility for connecting an optional external high stability local oscillator. If this is derived from a GPS or rubidium source then the receive converter will have exceptional frequency accuracy, as required for some SDR applications. In addition, the internal oscillator is accessible in order to allow it to be used as the LO for an accompanying transmit converter or with a second converter for use in a dual channel

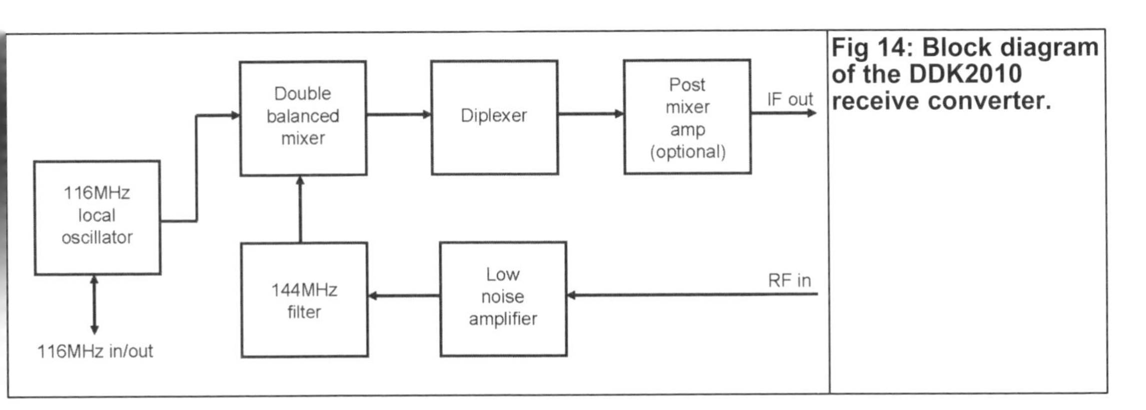

Fig 14: Block diagram of the DDK2010 receive converter.

synchronous receive system suitable for EME polarisation-dependant reception or simple diversity reception. It is also possible to use this output to feed into an external Reflock unit [5]. The Reflock can then phase-lock the internal 116MHz crystal oscillator. A block diagram showing the various stages of the receive converter is shown in Fig 14.

Circuit description

The DDK2010 receive converter circuit schematic diagram is shown in Fig 15.

Local oscillator

A familiar and reliable two stage 'Butler' overtone crystal oscillator is used to generate a 116MHz local oscillator signal that is fed into the LO port of the ADE 1 (MX1) double balanced mixer.

Contrary to normal practice, the oscillator maintaining stage (TR1) uses a switching transistor. High gain, low noise, transistors, can cause stability problems in this stage. I spent a great deal of time experimenting with different transistors in the Butler two stage overtone crystal oscillator and found that the BSV52 gave consistent and stable operation. Subsequent phase noise measurements (at Microwave Update 2009) confirmed the good performance of this arrangement. The second stage (TR2) is the oscillator limiter, biased for soft limiting so as not to seriously impact phase noise performance. This arrangement was chosen for simplicity over the more usual dual diode limiter. The oscillator output is untuned and delivers about +6dBm to the following mixer stage.

Since the first stage is basically a grounded base amplifier its emitter input impedance is well defined, as 26/Ie (where Ie is the TR1 emitter current in mA). In normal operation this gives an input impedance of approximately 7Ω. By incorporating a series 43Ω resistor between the external input and the emitter of TR1, an excellent match to 50Ω will be obtained at the external IN/OUT connector. In practice, 39Ω can be used with little change in performance. The required input level is -6 to +6dBm. This same connector can alternatively be use as an output, to take a sample of the internal oscillator for use in either an accompanying transmit converter or second receive converter. In both cases the series resistor should be increased in value to minimise interaction with the oscillator, and then a buffer amplifier may be required. A suggested starting value for R14 is 510Ω when the port is used as an output.

Mixer and diplexer

A commercial double balanced diode ring mixer type ADE-1 from Minicircuits Laboratories is

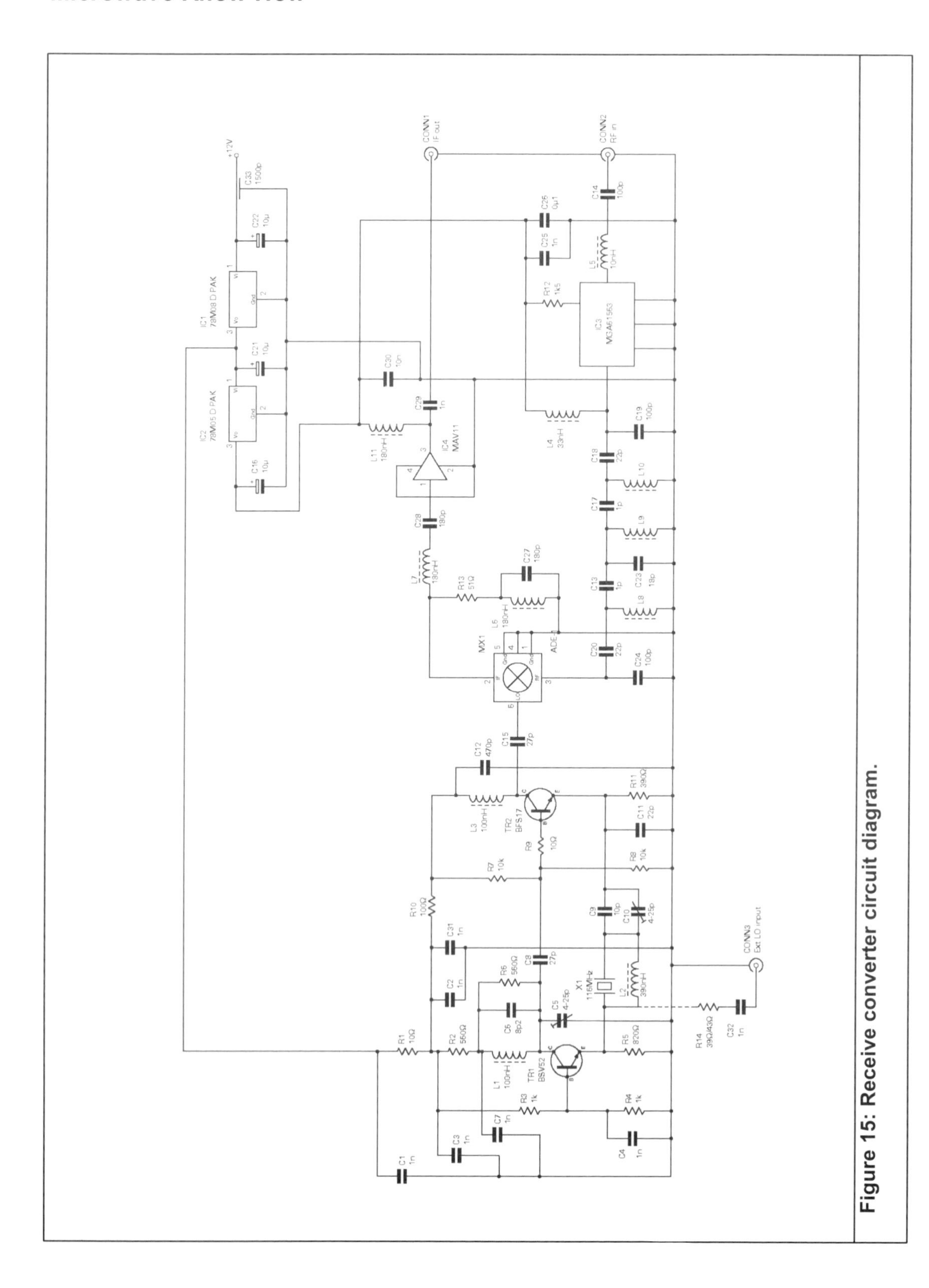

Figure 15: Receive converter circuit diagram.

used in the DDK2010. Although this is a standard level 7 mixer (i.e. requiring an LO input at +7dBm), it is slightly under-run in this design because of the low output from the 'Butler' oscillator. If this is a problem then the 8V regulator can be changed to a 10V device, which will cause TR2 to deliver slightly more output to the mixer.

A 28MHz diplexer consisting of R13, L6, C27, L7 and C28 terminates the mixer IF port in a good 50Ω match at all relevant frequencies. In practice this improves the conversion loss of the mixer and reduces the level of unwanted mixing and injection frequencies appearing at the IF output.

RF and IF amplifier stages

An Avago MGA61563 [6] was chosen for the RF stage because of its good noise figure at 144MHz, high dynamic range and excellent stability. This amplifier provides about 20dB of gain at 144MHz. The MGA61563 is a GaASFET MMIC that is useable from below 100MHz to over 6GHz. Its dynamic range can be altered by the amount of bias current the device is programmed to draw. Resistor R12 sets the bias current to about 45mA from the 5V supply, which provides a high dynamic range whilst maintaining an acceptable total current draw for the whole converter. If for example the converter is to be battery powered, R12 can be changed to 3.9kΩ. This will reduce the current to approximately 22mA, but with a consequent reduction in dynamic range.

IC4 is a MAV11 silicon, broadband, low noise post-mixer amplifier. It is optional, depending on whether you need the extra 12dB of IF gain it provides. If it is not required, leave out L11 and strap across the input and output of the IC4 stage. Another possible reason to omit IC4 is to save a further 40mA.

Band pass filter

The 144MHz inter-stage band pass filter consists of three tuned circuits, top capacitively coupled, with capacitive tapping for 50Ω input and output impedance. The shielded inductors are tuned with aluminium cores. 5mm square Coilcraft 165 series coils were chosen because they are currently available. Some other candidate coils are now obsolescent or on long delivery times.

Power supply section

The oscillator section gets its supply from 8V regulator IC1, which also supplies the 5V regulator, IC2. The output from the regulator is already well filtered, but it is possible to improve the phase noise performance by a few extra dB by connecting a 100µF aluminium electrolytic capacitor across C21.

The 5V regulator supplies the MGA61563 and the MAV11 post mixer amplifier. As arranged, the MAV11 is slightly current-starved, because it should bias to 5.5V on its output pin. In practice the 5V limitation does not seem to affect the gain or noise figure, although its dynamic range may be slightly reduced. If this is a concern then the MAV11 could be powered directly from the 8V regulator but it will need an additional wire strap from the 8V regulator and a 43Ω current limiting resistor in series with L11. Use a 1206 size resistor due to the high power dissipation.

By using a 'low drop out' (LDO) regulator for IC1, the converter could be powered from a 9V battery if required, since the LDO regulator will function down to about 8.5V input.

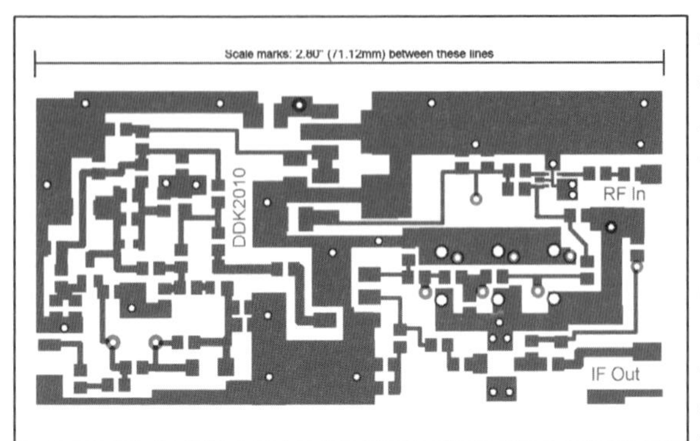

Fig 16: PCB layout for the DDK2010, reproduced at close to 1:1. Scale marks are shown to allow exact scaling

Construction

The receive converter uses an etched, 1.6mm thick, FR4, double sided printed circuit board. The PCB layout is shown in Fig 16 as close to 1:1 as possible but it is marked with a reference size to aid correct scaling. The parts list is shown in Table 2 and the component layout is shown in Fig 17. The board has a continuous ground plane on the other side. All of the surface mount components go on the track side. Only the crystal, the three adjustable coils and the single insulated wire link are mounted on the ground plane side of the board.

0805 size surface mount parts are used wherever possible, but the pad spacing is such that 0603 size parts could also be used. Low value 0805 size capacitors and fixed inductors, often used in RF designs, are now getting harder to find. Both regulators are rated 500mA and are available in the common DPAK package.

Suitable SMD parts are available in the UK from many suppliers including Farnell, RS and Rapid Electronics [7]. Specialist component part numbers are shown in the parts list. Since there are a large number of alternatives available for many of the other parts it would be impractical to list them all. The adjustable coils are only available in the UK from Coilcraft Europe Ltd [8]. Possible 116MHz crystal suppliers are listed [9].

The board is designed to be soldered into a 37mm x 74mm x 30mm size Schuberth tinplate box to provide screening from external signal pickup. These boxes are obtainable from [10].

- The PCB should be cut to 71.6mm x 34.5mm in order to be a tight fit into the box. File small notches in the two diagonally opposite corners to clear the overlapping flanges in the box. Do not solder the box together at this stage.

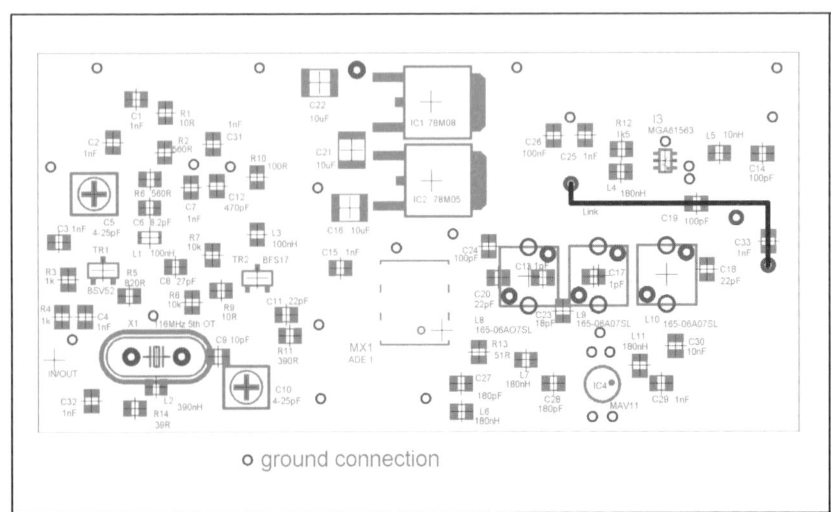

Fig 17: Component layout for the DDK2010,

Table 2: Parts list for the DDK2010,

Resistors	
R1, 9	10Ω
R2, 6	560Ω
R3, 4	1kΩ
R5	820Ω
R7, 8	10kΩ
R10	100Ω
R11	390Ω
R12	1.5kΩ
R13	51Ω
R14	39 or 47Ω
Capacitors	
C1, 2, 3, 4, C7, C25, C29, C31, C32	1nF
C5, 10	4 - 25pF Trimmer
C6	8.2pF
C8, C15	27pF
C9	10pF
C11, C18, C20	22pF
C12	470pF
C13, C17	1pF
C14, C19, C24	100pF
C16, C21, C22	10µF, 16V
C23	18pF
C26	100nF
C27, C28	180pF
C30	10nF
C33	1500pF Tusonix feedthrough capacitor

Inductors	
L1, L3	100nH
L2	390nH (may not be required)
L4	33nH
L5	10nH
L6, L7, L11	180nH
L8, L9, L10	Coilcraft 165-06A07SL
Semiconductors	
IC1	78M08 (500mA)
IC2	78M05
IC3	Avago MGA61563
IC4	MAV11
TR1	BSV52
TR2	BFS17
MX1	ADE-1
Miscellaneous	
X1	116MHz 3rd Overtone crystal, HC43/U
Connector 1, 2, 3	SMA 2 hole female bulkhead connectors
Box	Schuberth 34 x 74 x 30mm

- Prepare the PCB by drilling the various through-board holes. The drill size should be no larger than 0.6mm diameter, except for the lugs for the three adjustable coils. These holes should be drilled to 1mm diameter. Solder small-gauge tinned wire through the ground holes. To avoid unwanted short circuits, ensure that the top side (ground plane) copper is cleared around the holes for the pins of the adjustable coils and the crystal as well as the wire link holes and the regulator DC input. A hand-held 3mm twist drill works well.
- Solder all SMD parts onto the PCB using a fine-tipped soldering iron and fine gauge pre-fluxed solder. Start with the resistors and capacitors, followed by the inductors and finally the transistors, integrated circuits and regulators. The regulator 'tabs' should be soldered to the PCB ground plane for heatsinking.
- Prepare and solder the insulated single wire link on the ground plane side of the PCB.

- Finally, solder the adjustable coils and the crystal to the board, ensuring they are fitted on the ground plane side of the PCB. The crystal should be pushed down snugly onto the ground plane, as should the three coil shielding cans. The coil cans should be soldered to the ground plane along their lower edge.

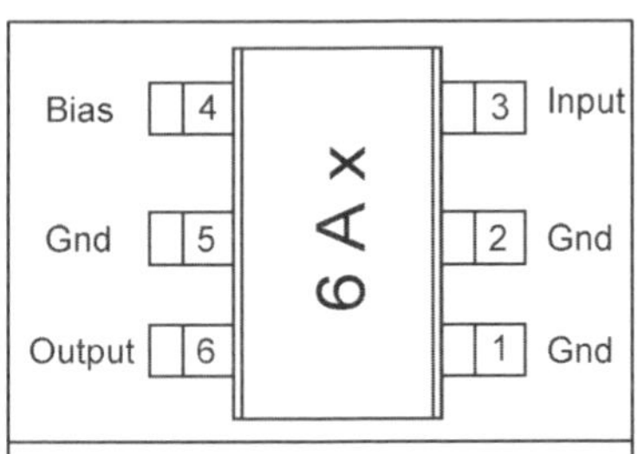

Fig 18: Pinout of the MGA61563. 6A is the device part code and x is the date code.

The most difficult part to mount is the MGA61563, IC3. This device uses a miniature 6 pin RF SOT363 package. Before soldering, ensure that the package is correctly oriented, with the input and output pins in the correct position. See Fig 18 for pinout information. IC3 must be carefully soldered so that its 6 leads are just resting on the 6 PCB pads. It is probably best to initially tack solder one lead, then carefully manoeuvre the package with a pair of tweezers so that the remaining 5 leads lay flat to their respective pads. Take care not to break off any of the 6 leads from the package. Solder quickly, using a small soldering iron and very fine gauge solder – 28 SWG is recommended.

Mark the inside of the box so that the PCB ground plane will be positioned far enough below the rim of the box that the height of the crystal and the adjustable coils can be accommodated without protruding beyond the rim of the box. Also, and most important, mark the correct side of the box so that the corner notches in the PCB are on the right side to engage with the box overlapping flanges. It is easy to get this wrong!

Select and prepare three 2-hole SMA connectors by removing all but 0.5mm of the Teflon™ insulation from the spill of the connector. Mark the position of the three SMA connectors on the inside of the box and drill appropriate size holes (usually 4mm diameter) so that the insulation of the connector just protrudes into the box, but not so far as to interfere with the PCB. The holes for the connectors should be positioned so that the connector spills lay flat onto their respective connector pads. It is advisable to mark and drill holes for the three connectors at this stage as it will be found very difficult to do so after the PCB is mounted in the box, should you change your mind about needing the input/output socket at some later stage. Do not solder the connectors to the box at this stage.

Mark and drill a hole for the supply feedthrough capacitor. If using a Tusonix solder-in feedthrough, you will need a 2mm diameter hole in the box. Do not solder the feedthrough into the box at this stage.

Assembling the box

In order to ensure the best fit of the PCB into the box use the following procedure:

- Solder one edge of the PCB into one half of the box by tack-soldering the ground plane side of the board. Ensure the corner notches are on the correct side. Now carefully fit the PCB and its side of the box into the lid of the box. Jig the remaining box side into place inside the lid. The PCB, two sides and the lid should fit comfortably together.

- Place the top lid on the assembly and check that it also fits comfortably.

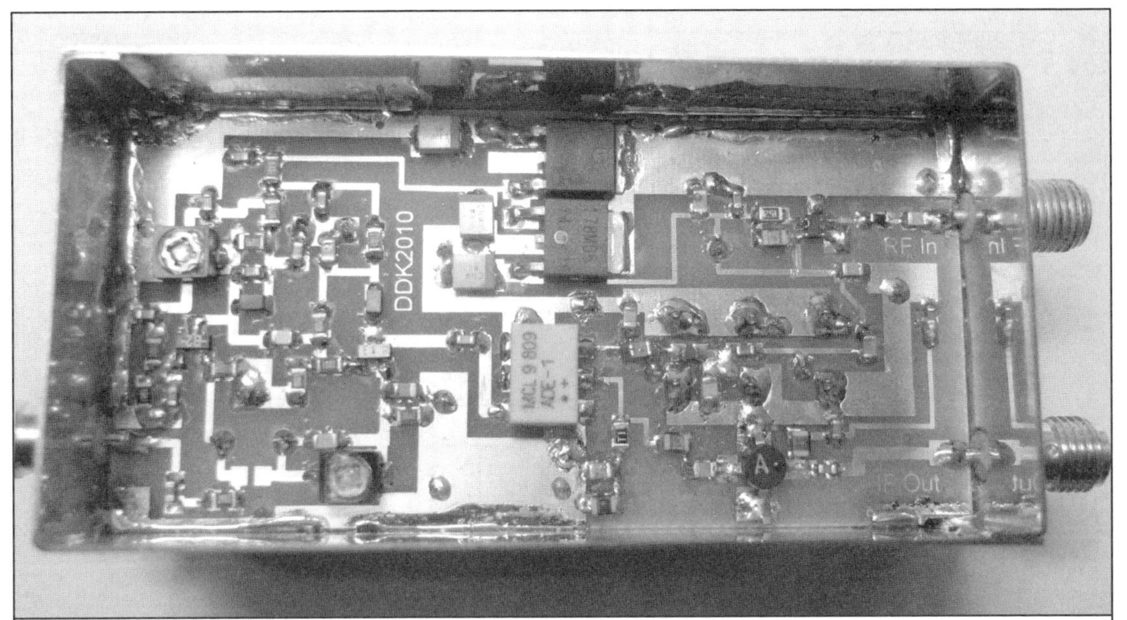

Fig 19: Component side of the DDK2010 144MHz to 28MHz converter.

- Remove the lower lid. Solder along the overlapping flanges of the box, forming a rigid box section.
- Replace the lower lid and remove the top lid. Seam-solder the remaining exposed parts of the box flanges.
- Having checked the PCB is level in the box, solder the previously unsoldered side of the PCB to the 2nd box side and then check that the lower lid still fits comfortably in place.
- When you are happy that all is well, go back and complete the seam-soldering of the PCB to the inside of the box. This will need to be soldered on both the ground plane side of the PCB and along some of the track side.
- Solder the connectors and the feedthrough capacitor on the outside of the box, making sure that you include a 2mm hole diameter solder tag under the feedthrough before soldering it in place. The lack of a suitable supply grounding point can be an annoyance when testing!
- Solder the connector spills to their respective PCB pads.
- Connect a short insulated wire from the feedthrough capacitor to the regulator input.

Fig 19 and 20 show the top and bottom of the completed converter.

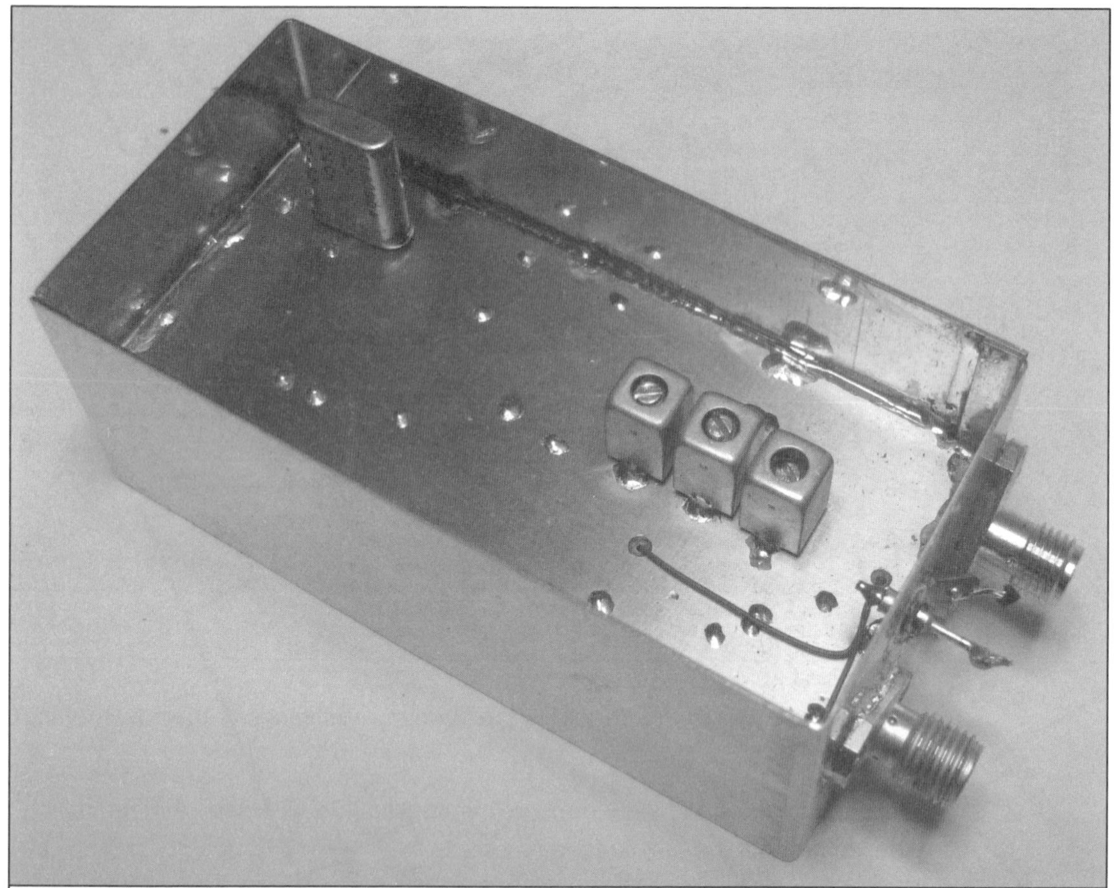

Fig 20: Ground plane side of the PCB, showing the 116MHz crystal and the band pass filter coils.

Alignment

Before connecting any power, check with an ohmmeter that there is no short between the supply input feedthrough capacitor and ground. If all is well, connect a +12V current limited power supply set to less than 150mA to the feedthrough capacitor and 0 volts to ground. Check the total current taken, which should be no more than about 100mA, depending on whether IC4 has been fitted. If it is higher, check for faults such as short circuits, incorrectly placed components or reversed polarity tantalum capacitors. If all is well then proceed to set up the crystal oscillator. This is easily done by listening to the output from a receiver tuned to 116MHz. Most scanners and aircraft band receivers will cover this frequency. Use FM or AM to listen for a distinct change of noise from the receiver when C10 is adjusted. C5 will need to be set near minimum capacitance to allow the oscillator to operate at 116MHz. If the receiver is within a few metres of the converter the noise change will be very apparent. The final local oscillator frequency can be set more accurately later. The Butler oscillator is a very reliable starter. This version, with the BSV52 first stage, is particularly docile. Even very lazy crystals should work well in this circuit.

If you have access to a spectrum analyser and signal generator, then you will probably already know how to set up the adjustable coils of the band pass filter by adjusting for maximum IF output at 28MHz with the signal generator set to 144MHz. Do not set the signal generator to greater than -20dBm output, to avoid saturating the converter output.

If you do not have access to test equipment then connect a HF receiver, tuned to the 28MHz band, to the IF output. Connect a 144MHz antenna to the RF input. Tune the receiver to a strong local repeater output channel or to a local beacon of known frequency. Assuming a signal can be heard, adjust L8, L9 and L10 for maximum signal. If no signal can be heard, you should still hear a small but noticeable increase in noise output as the 144MHz bandpass filter is adjusted to cover the correct frequency range. The limited adjustment range of the coils in the filter should prevent any chance of mistakenly tuning to the image at 88MHz. When correctly adjusted, the band pass filter will exhibit a nearly flat frequency response across the 144 to 146MHz frequency range.

If you can now hear a beacon or repeater, but it is slightly off frequency, adjust C10 to obtain zero beat when using the receiver on USB (or LSB) mode. It may be necessary to slightly adjust C5 to bring C10 within range. In the event that you cannot move the frequency of the oscillator onto exactly 116.000MHz it may be necessary to connect the optional inductor L2 across the crystal. This is probably not required, but may help. If you are not within range of a repeater or beacon use another, suitably accurate, signal source to adjust the local oscillator frequency.

There are no other adjustments to make and the converter is now ready for use.

Performance

Table 3: Prototype receive converter performance.

Noise figure	better than 3.0dB
Gain	17.5dB
Input 1dB compression point	-16dBm
Bandwidth	3.0MHz
Image response	Greater than -80dB
Current consumption	100mA at 12V

The measured performance of the prototype converter is shown in Table 3. Note that the alignment of the bandpass filter, whilst not critical, must be done carefully as small changes in filter loss, due to poor alignment, can result in an unacceptably high noise figure in such a low gain converter design. The overall noise figure of the converter is lower (better) without the MAV11 stage. This may not sound very intuitive but is due to the 'second stage' noise contribution of the MAV11 stage. The same degradation in noise figure will be experienced when the converter is fed into a typical 28MHz superhet receiver or SDR receiver.

Extending the use of the receive converter

The provision of an external local oscillator input/output makes the DDK2010 very flexible since it is possible to connect an external, high stability 116MHz source in place of the internal crystal oscillator. If this is GPS-locked then the receive converter can become part of a completely frequency-locked receiver/transverter system for any of the bands on or above 144MHz. Alternatively, the internal 116MHz signal can be extracted from the input/output socket and used to feed into an accompanying transmit converter to produce a high performance transverter for 144MHz.

For 144MHz moon bounce (EME) enthusiasts, the 116MHz output can be fed into the corresponding input/output socket of a second DDK2010 receive converter (keep R14 as 39Ω or 43Ω) to produce a dual RF input system for use with a polarity sensing receiver that uses one converter on the horizontal antenna array and a second converter on the vertical one.

It is also possible to replace the MAV11 post mixer amplifier with a higher gain device such as the MAR6. The receive converter will have a similar noise figure with a gain up to 8dB higher, but the dynamic range will be reduced. An additional 91Ω resistor will be required in series with L11 to obtain the correct bias for the MAR6.

References

[1] Digi-Key Corporation - www.digikey.com

[2] Ed Johnson, AD5MQ web site - http://www.sdc.org/~edjohn

[3] Surplus Sales of Nebraska - http://www.surplussales.com

[4] Reflock - http://gref.cfn.ist.utl.pt/cupido/reflock.html

[5] Article by Sam Jewell, G4DDK, from RadCom March 2010

[6] SMD parts are available from many suppliers including www.farnell.com, http://uk.rs-online.com/web and www.rapidonline.com

[7] The Avago MGA 61563 data sheet can be downloaded from www.avagotech.com/docs/AV02-1471EN

[8] The adjustable coils can be obtained from www.coilcraft.com/general/sales_eu.cfm

[9] 116MHz crystals can be sourced from www.quartslab.com or www.eisch-electronic.com

[10] Schuberth Tinplate Boxes are sold by www.alan.melia.btinternet.co.uk

Chapter 7

Oscillators

In this chapter :

- DDS using the AD9951
- MMIC oscillator experiments

Nearly every amateur radio project needs a local oscillator. The days of the crystal oscillator and transistor multiplier chain are numbered. This chapter show two different approaches using modern integrated circuits.

DDS using the AD9951 [1]

Circuit technology is progressing with new ranges of Integrated circuits. The AD9951 DDS is far superior to a 14 bits digital to analogue converters in earlier direct digital synthesisers (DDS) like the AD9850 and AD9851 this is because it has a clock frequency up to 400MHz. This article describes a complete circuit using modern components.

Inspired by different publications and various digital frequency synthesis techniques [6 -10] I decided to attempt my own design. After a thorough study of the data sheet [11] the actual work began on this modern direct digital synthesiser. The first prototype that was developed with an external clock showed the performance that could be achieved. This approached the data sheet values, so could this be improved with circuit changes?

Circuit description

The circuit of this Direct Digital Synthesiser (DDS) project consists of four modules.

The block diagram is shown in Fig 1 the modules are:

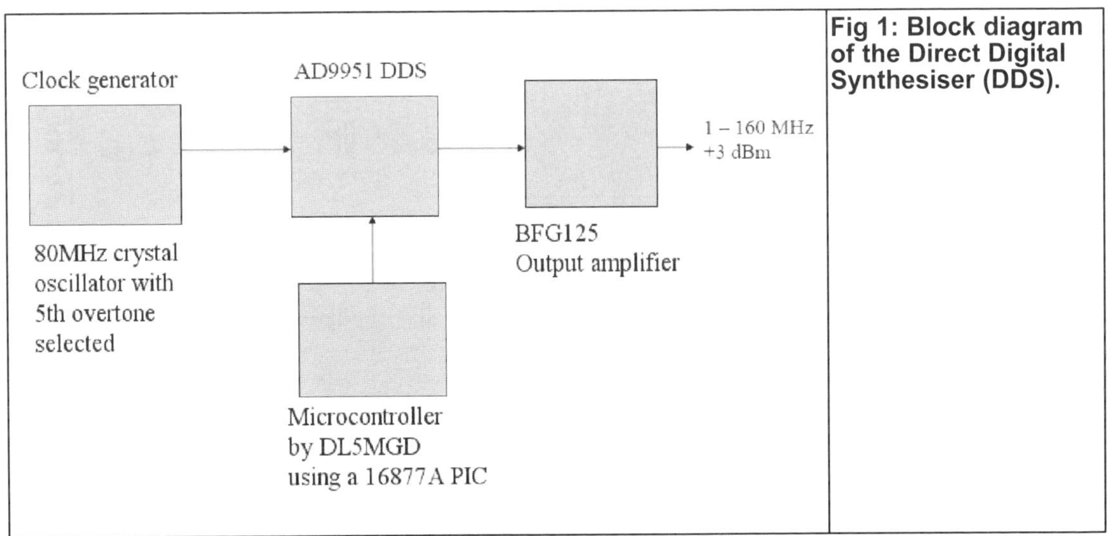

Fig 1: Block diagram of the Direct Digital Synthesiser (DDS).

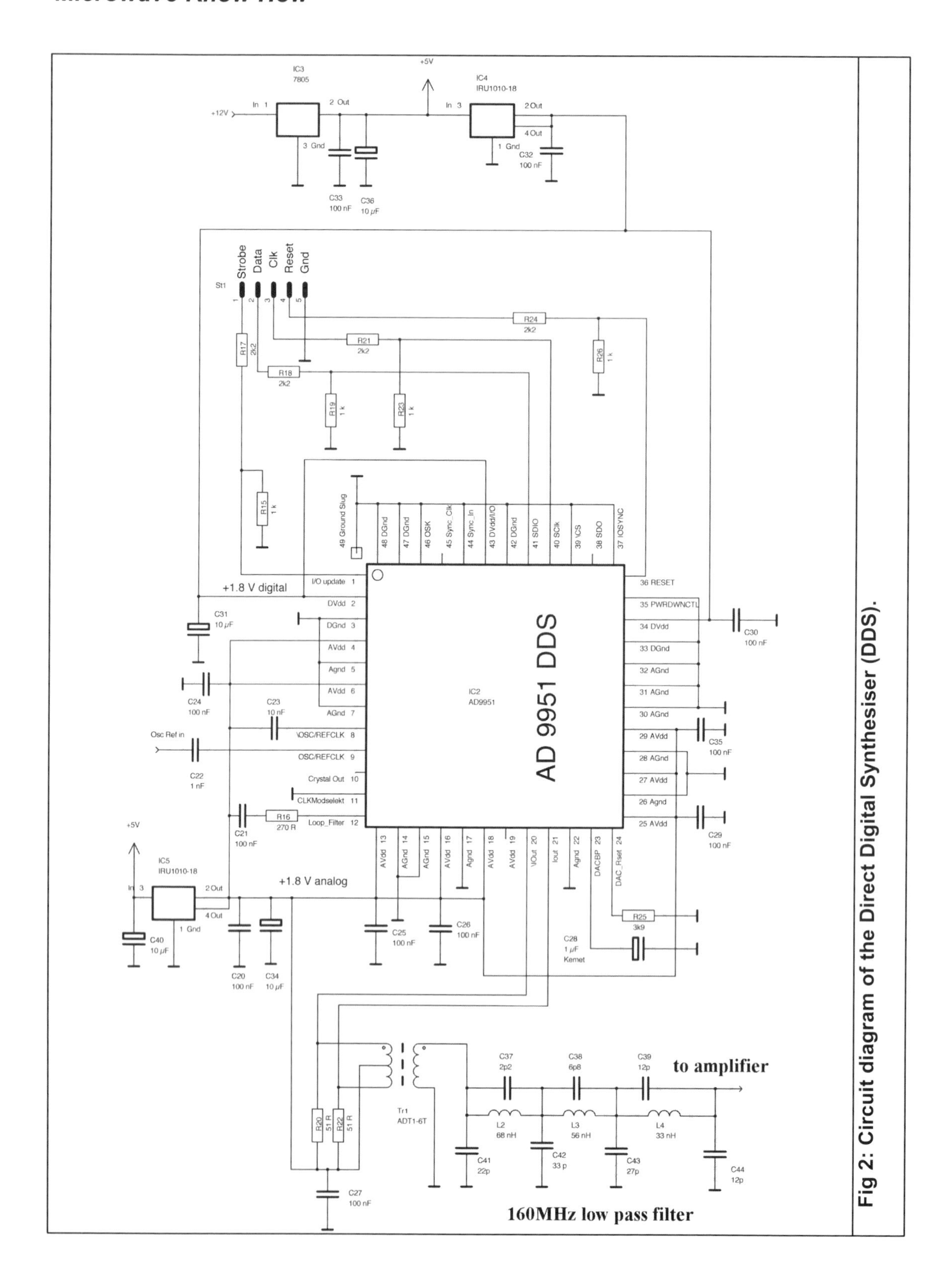

Fig 2: Circuit diagram of the Direct Digital Synthesiser (DDS).

- DDS
- 400MHz clock pulse generator
- Microcontroller to control the DDS
- Output amplifier

DDS

The circuit diagram of the DDS module is shown in Fig 2.

The heart of the DDS printed circuit board is the AD9951 DDS integrated circuit from Analog Devices. It is a chip in a rectangular package (Thin Quad Flat Pack - TQFP/EP) with 48 legs on a 0.5mm pitch and a relatively large soldering surface (thermal pad) under the IC to dissipated the heat generated. Commercially manufactured double sided and plated through printed circuit boards were used for the prototype because the tracks are only 0.2mm wide. The soldering area for the thermal pad is connected to the earth surface by plated through holes.

Voltage supply

The supply voltages of 1.8V for the digital and analogue part of the DDS IC are produced using two IRU 1010-18 linear voltage regulators fed from the stabilised 5V supply as shown in [6]. The data sheet [14] shows that these can give a regulated supply with a maximum current of 1A and a maximum input voltage of 7V. According to the data sheet the operating power used by the DDS is typically 161mW, i.e. a current 89mA. Taking the quiescent current of the two linear regulators into account using a 5V supply [13] gives:

445mW - 161mW = 284mW.

Switched mode controllers, as suggested in [5], would improve the power dissipation performance and that may be meaningful for QRP equipment but it can produce unwanted spurious products.

The inputs for programming the internal DDS internal are designed for input voltages of 3.3V maximum, if pin 43 of the chip (DVDD_IO) is supplied with 3.3V. Because the microcontroller with has a supply of 5V a third voltage regulator (5V to 3.3V) could be used for DVDD IO or the high level from the microcontroller can be reduced using voltage dividers. Therefore DVDD IO is connected to the 1.8V voltage regulator for the digital section. The digital signals; Clock, Data, Strobe and Reset from the external microcontroller operated with 5V are connected to the DDS data inputs by 2.2k/1kΩ resistive voltage dividers.

Output connection

The Digital to Analogue Converter (DAC) outputs of the DDS are complimentary current outputs. An external resistor connected between the Rset pin (24) and ground controls the full-scale current. The value of the resistor is calculated as follows:

Rset = 39.19/Iout [Rset in kΩ, Iout in mA].

The best spurious free dynamic range is achieved with a maximum current of 10mA, which means a resistor of 3.9kΩ. From the data sheet the operating range of the DAC outputs is from the analogue supply voltage (AVDD) - 0.5V to AVDD + 0.5V for a supply voltage of 1.8 V giving the limits 1.3V to 2.3V, the maximum voltage swing is therefore 1Vpp = $0.35V_{eff}$. The two DAC outputs are current sources capable of supplying 1 to 10mA.

With these conditions the voltage developed across the load resistance can be between 1.8V and 1.3V relative to AVDD because the voltage rises by 0.5Vpp. Under these conditions can a 50Ω load resistance be used? Voltages outside of this range can cause substantial damage and can permanently damage the DAC.

The DAC signals should be coupled using a centre-tapped transformer feeding the load resistance as described in [11]. The resistance value is selected in such a way that the limiting values indicated above are not exceeded. The centre tap is connected to AVDD so that the DAC see a load resistance of 25Ω with a 50Ω load on the transformer. This method of connecting the DAC output has the following advantages according to [11]:

- Disturbances on supply voltage AVDD such as remnants of the clock pulse and other unwanted signals are effectively suppressed with a high symmetry transformer [11, page 48].
- The output voltage is doubled compared to using one DAC output.
- Reflections from a low pass filter following the DDS are effectively suppressed [11, page 47]

It is amazing that no transformers are used in the many circuits published for the well known AD9850 DDS. A fast operational amplifier, connected as a sum and difference amplifier, is shown in the circuit given in [4]. I0CG uses an unknown type of transformer from Minicircuit in [8, 9].

For the first attempts an ADT1-6 transformer was used, it is readily available and is suitable for a range from 10kHz to only 125MHz. In order to use the upper frequency range of the DDS the better transformer type T1-1T (80kHz to 200MHz) should be used. Both transformers have are 1: 1 but the T1-1T has a different housing. The voltage at the output of the transformer is twice the voltage across one half of the primary coils; this can be easily proved by measurement with an oscilloscope.

The latest implementation by Martein Bakker, PA3AKE [17] uses a capacitor (C28) on the DAC baseline-decoupling pin (DACBP pin 23) of the DDS. It filters the amplitude noise produced by the internal Band gap reference that would otherwise modulate the DAC output. This pin is no longer present on new DDS components e.g. AD9912.

Low-pass filter

The seven pole elliptical low pass filter circuit and component values were taken from the Analog Devices evaluation board [13].

Output amplifier

The DDS supplies an output level of approximately 6dBm that is sufficient for many applications. Nevertheless a higher output level is desirable. Therefore an output stage was designed with the circuit diagram shown in Fig 3.

To maintain the harmonic suppression of approximately 60dB obtained from the DDS a linear amplifier was needed. Experience with MMIC amplifiers like the ERA 3 showed that these would not give sufficient harmonic performance. A test with a BFR96 amplifier had undesirable results;

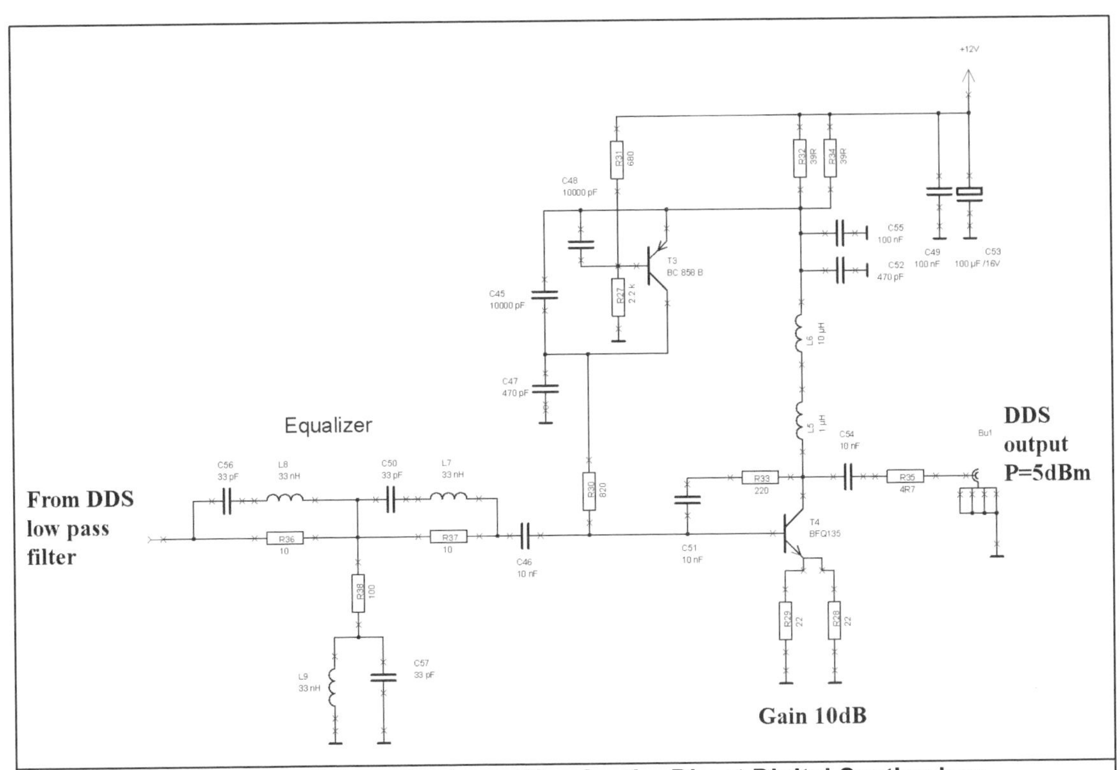

Fig 3: Circuit diagram of the output amplifier for the Direct Digital Synthesiser.

it suffered from wild oscillation in the UHF range and damaged the transistor.

A broadband amplifier using a BFG135 with negative feedback was developed and designed for stable gain of 10dB. With a current of 80mA it achieved the required 60dB harmonic suppression. The negative feedback circuit was designed with the commercial program "Genesys" version 8.1 (an Eagle product, now Agilent) with the simulation circuit shown in Fig 4. No DC voltage operating points can be examined with the basic version of the program that the author used so no operating point stabilisation was included in the simulation.

My first idea was to use a frequency varying negative feedback in order to compensate for the gain loss of the DDS output. This was done using a 100nH inductor in series with the feedback resistor and capacitor from the collector to base and a capacitor from emitter to ground. This circuitry seemed to work for the simulated frequency range of up to 200MHz. As the BFG135 has an ft of 6GHz the simulation frequency range must be extended in order to see if the circuitry is unconditionally stable (stability factor $k > 1$) even at higher frequencies. By doing so the simulation showed that for frequencies above 500MHz k is < 1. Therefore this compensation scheme is not feasible. By inspection of the test circuitry in the data sheet of the BFG135 [18] the feedback circuitry was changed (L2 set to 22nH, the cap from emitter to ground omitted and a small (10Ω) series resistor from the collector to the output included, see Fig 4. This gives a stable gain and stability factor over the frequency range up to 1GHz, see Fig 5.

In order to compensate for the sin c function, a bridged T-filter was designed. This filter is put

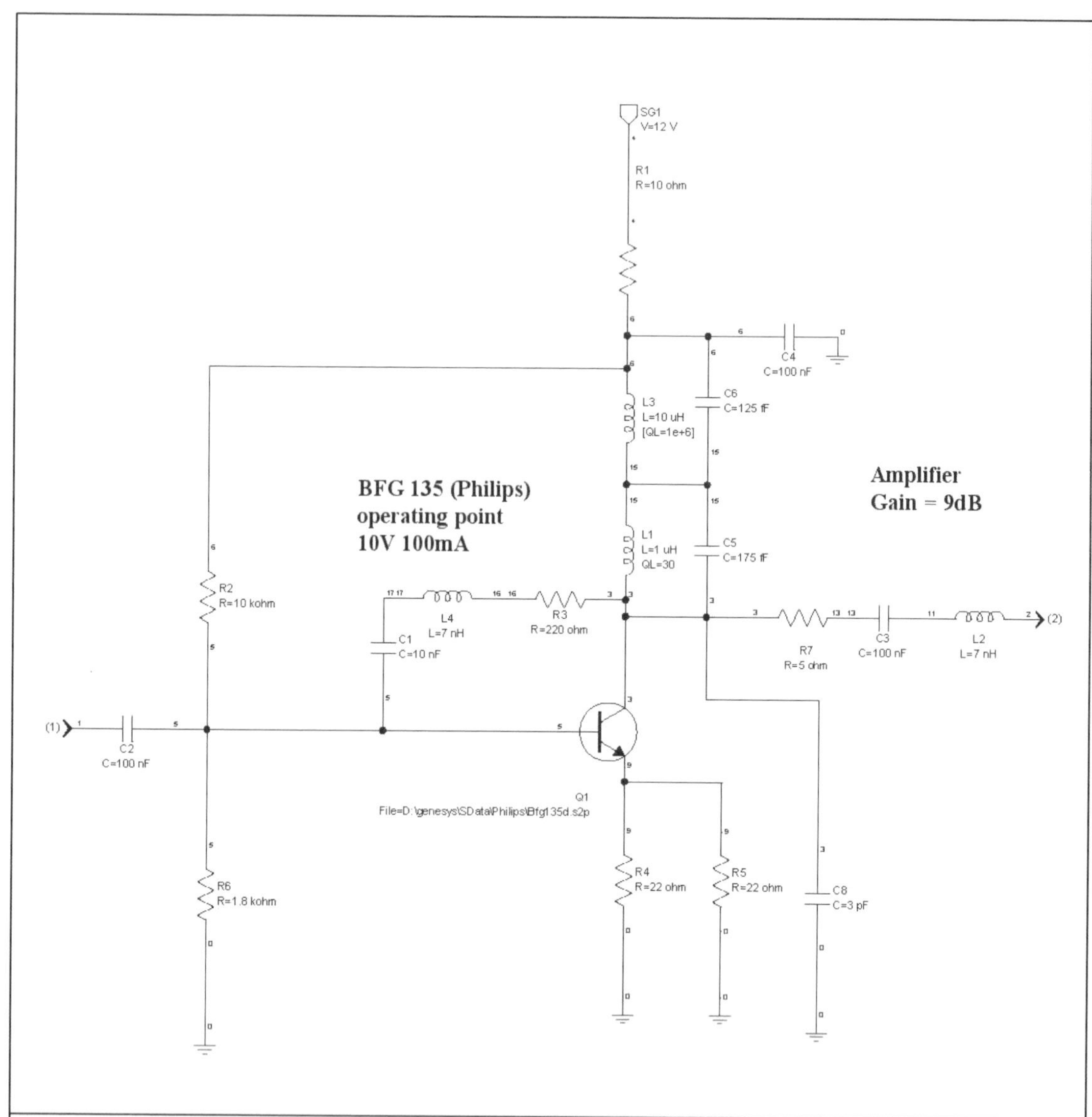

Fig 4: The simulation circuit of the output amplifier developed for use with Gensys.

between the lowpass filter of the DDS output and the input of the output amplifier. This circuitry gives a constant input and output impedance (like a diplexer at the output of a mixer). The attenuation is 4dB at low frequencies but as the attenuator is bridged at about 160MHz there is nearly no attenuation at higher frequencies. See Fig 6 that shows the frequency response of cascaded bridged T-filter and output amplifier. This circuitry is not implemented in the author's test PCB but is included in the updated PCB shown in this article.

The simulation up to 1GHz is shown in Fig 6. The gain drops 500MHz drops so that no self-oscillation is expected.

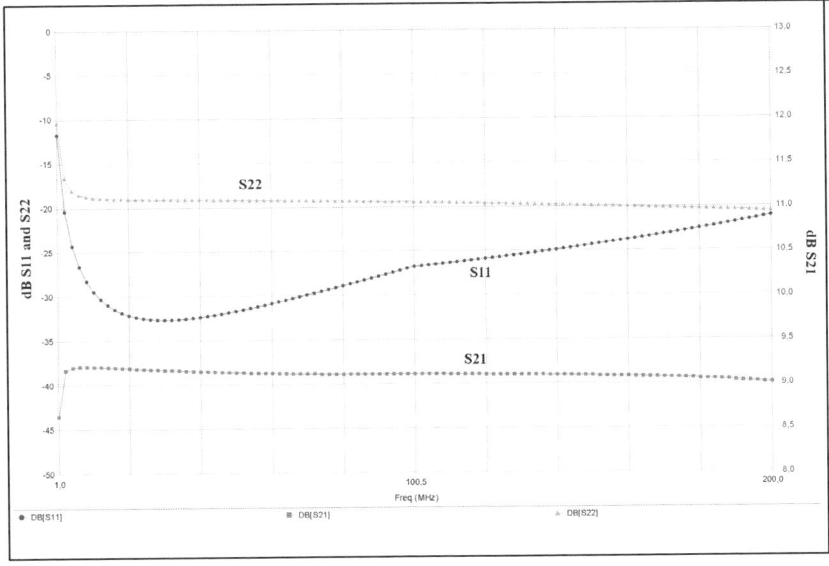

Fig 5: The simulation result showing the S parameters S11, S22 and S21 for the output amplifier in the frequency range from 1 to 300MHz.

Clock production

The circuit diagram of the 80MHz crystal oscillator module is shown in Fig 7. Experiments with an external 20MHz clock and internal multiplication as shown in the data sheet [12] show that the best phase noise is obtained using an external 400MHz clock. The clock is generated from an 80MHz crystal oscillator with a TTL compatible output (Pout = +17dBm for the fundamental frequency). The fifth harmonic is selected using a two pole helix filter. The filter is fed via a coupling capacitor because the TTL oscillator output has a DC offset. The prototype gave a measured output of –6dBm at 400MHz see Fig 8. Any frequency offset can be compensated using the software in the microcontroller [6].

When selecting harmonics of an oscillator the phase noise increases by 20 log N, where N is the harmonic number selected. Selecting the fifth harmonic increases the phase noise by 20 log 5 =

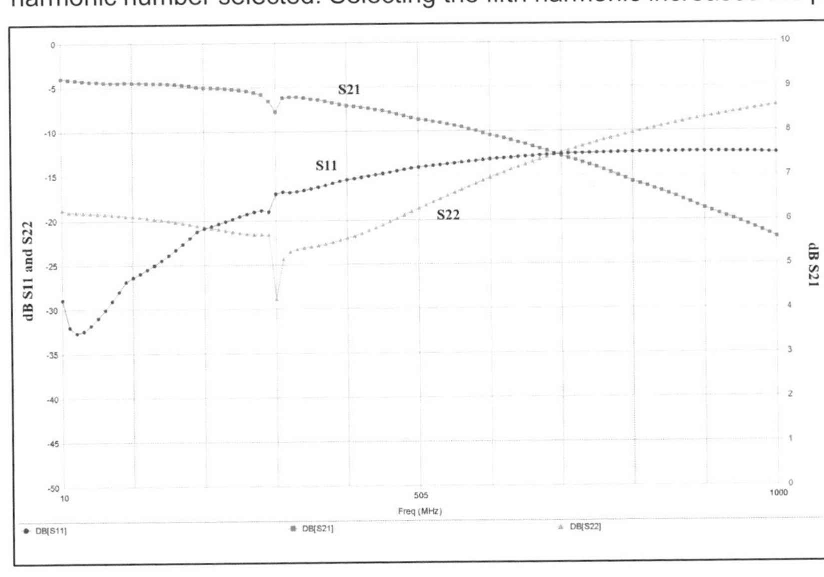

Fig 6: The simulation result showing the S parameters S11, S22 and S21 for the output amplifier in the frequency range up to 1GHz.

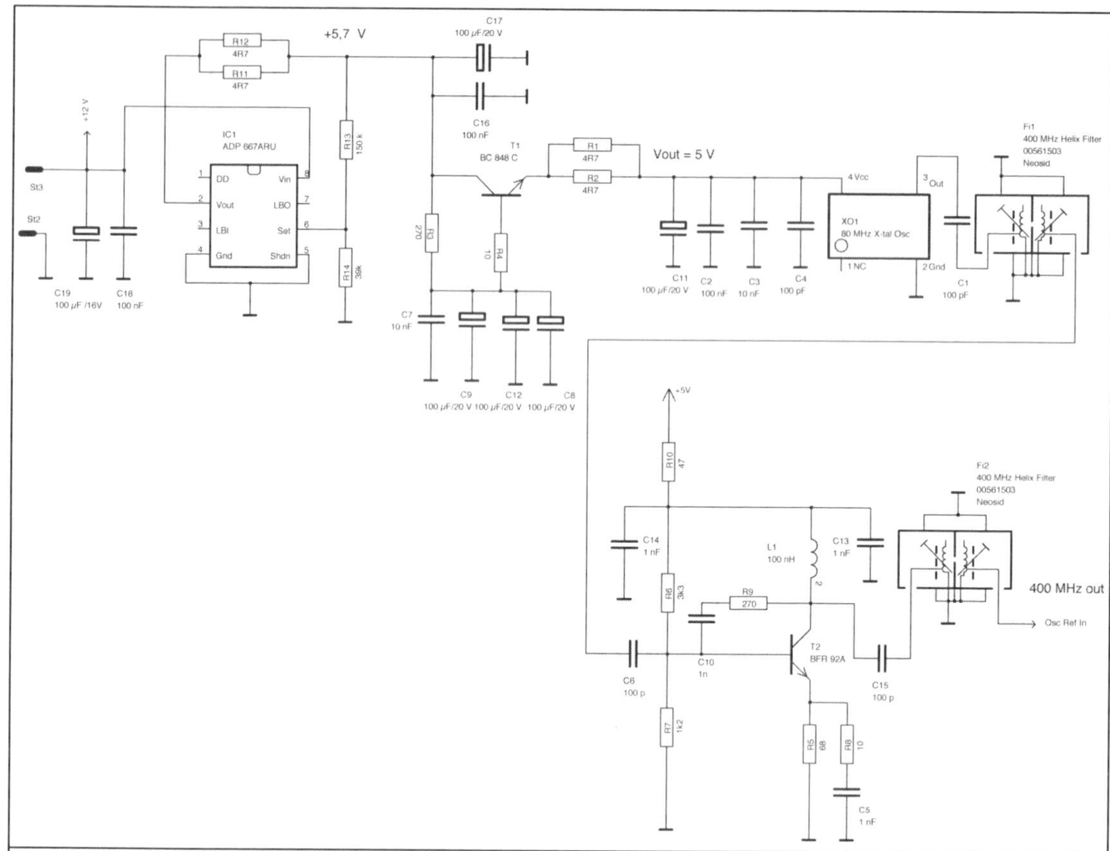

Fig 7: Circuit diagram of the 80MHz crystal oscillator module.

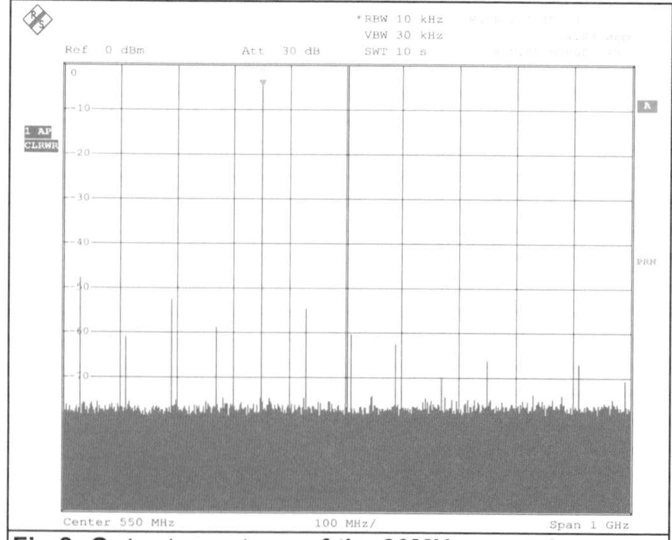

Fig 8: Output spectrum of the 80MHz crystal oscillator after a 400MHz two-pole Helix filter.

14dB for a fourth harmonic this would be 12dB. Attempts to select the fourth harmonic from a 100MHz square wave oscillator feeding a diode multiplier [15] failed. The fundamental could be filtered to give a sine wave to feed the diode multiplier [15] and the fourth harmonic selected using a low pass filter but the additional cost cannot be justified.

Fig 9 shows the measured phase noise of the crystal oscillator on the fundamental (80MHz) and fifth harmonic (400MHz). The degradation in phase noise was expected to be 14dB but was 20dB at 100kHz from

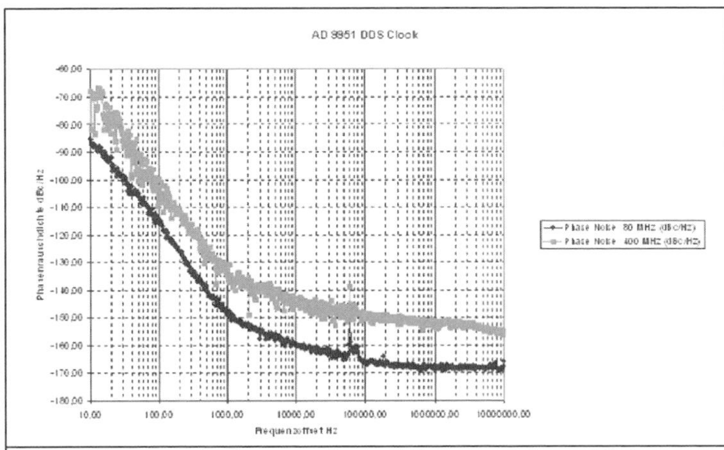

Fig 9: Measured phase noise of the crystal oscillator on the fundamental (80MHz) and the fifth harmonic (400MHz).

the carrier. Since these measurements were taken at different times and with different power supply smoothing the results are difficult to evaluate.

The phase noise of a 100MHz crystal oscillator was also measured and it was worse than the 80MHz oscillator at the fundamental frequency therefore because of the higher price of this oscillator the 80MHz oscillator is more favourable. The good phase noise of the cheap crystal oscillator was a surprise.

However it must be mentioned that the phase noise of the oscillator is sensitive to noisy supply voltages. Therefore the supply voltage for the oscillator uses a low-noise low drop out voltage regulator and a filter. The filter consists of an NPN transistor emitter follower with high value

Table 1: Modificatios to the source code that can be downloaded from [5].

```
RESET:
    Bsf      reset          ; Reset the AD9951, (high signal) so that all registers are in a defined condition
    call     wait_1ms       ; Reset goes "high "and remains there for the 1 ms specified in the subroutine
    bcf      reset          ; Reset back to "low "
; ------Initialization of the AD9951----------
    Init                    ; register cfr2 of the ad9951
    movlw    b `00000001 `  ; address cfr2 in the write mode
    movwf    daten          ; call ser_aus
    movlw    b `00000000 `  ; highest byte of the cfr2 (internally not used in the ad9951)
    movwf    daten
    call     ser_aus
    movlw    b `00000000 `  ; second highest byte of the cfr2
    movwf    daten
    call     ser_aus
    movlw    b `00000110 `  ; lowest byte of the cfr2 clock = 400 MHz,
                            ; internal multiplicator switched off
    movwf    daten
    call     ser_aus
    bsf      strobe
    bcf      strobe
    return
; --- END to initialization of the AD9951-----
```

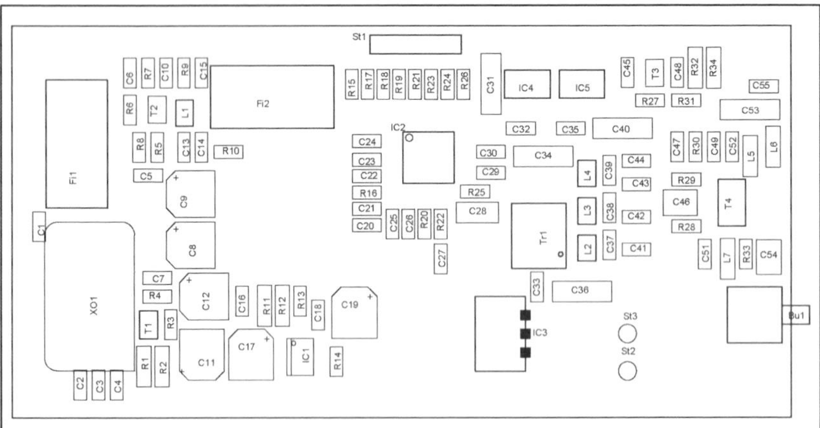

Fig 10: Component layout of the DDS PCB.

electrolytic capacitors connected from its base to earth. The effective value of these electrolytic capacitors is increased by the current amplification factor of the transistor. With a current amplification factor of 100 the emitter capacitance is 100 x 300µF = 0.3mF. With this high capacitance value any noise voltages are effectively suppressed.

A capcitively loaded emitter follower can oscillate; this can be reduced with the 10Ω resistor in the base circuit. A resistor in the emitter circuit reduces the starting current flowing into the electrolytic capacitors. The current drawn by the crystal oscillator in the prototype was 60mA so two 4.7Ω of resistors (R1, R2) in parallel were used to minimise the voltage drop.

The frequency stability with temperature and ageing of the cheap crystal oscillator has not been investigated.

The oscillator signal can be fed to the DDS either symmetrically using a Balun or asymmetrical [12]. The oscillator inputs are internally connected to 1.35V linked therefore a coupling capacitor is required. When asymmetrical feed is used, pin 8 must be decoupled to AVdd with a 0.1µF capacitor. The oscillator output, Crystal Out, pin 10 can be switched off by setting an internal register (CFS2>9> = high). The data sheet for the DDS does not specify the Balun that can be used for symmetrically feeding the clock. The input level of the externally clock can be between +3dBm (890mVpp) and -15dBm (112mVpp) A high level is said to improve the phase noised but this is not quantified. For the maximum of the clock level the voltage is 1.35V + (0.89V)/2 = 1.795Vs. Because the harmonic suppression was not sufficient at 400MHz a buffer amplifier with another two-pole helix filter was developed.

Microcontroller

The microcontroller developed by Andreas Stefan, DL5MGD [6] was used to control the DDS. The firmware was modified for the current circuit; the modifications to the source code that can be downloaded from [6] are shown in Table 1. The second bit in the second highest byte of CFR2 is set high (b `00000010 `) making the Crystal Out, pin 10, inactive possibly diving better harmonic performance.

The first attempt to make this circuit on strip board failed so a ready-made PCB was purchased from Andreas and it worked first time. The circuit diagram of the microcontroller, operating instruction in English and the tiff files to make the single sided PCB is available from Andreas web site [6].

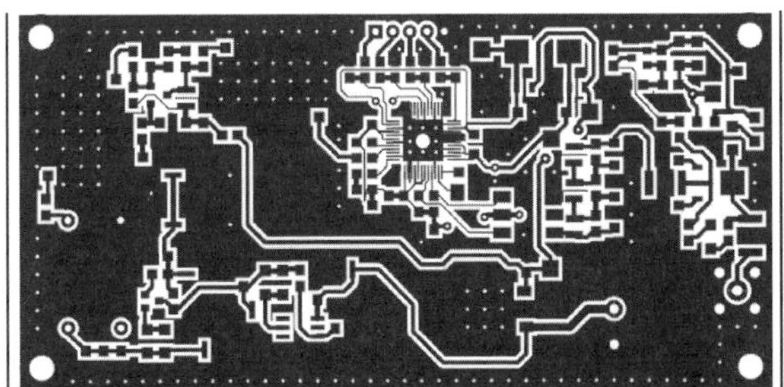

Fig 11: SMD component side of the DDS PCB.

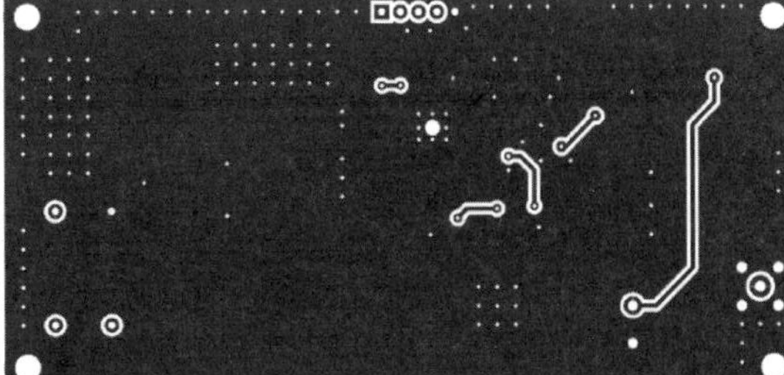

Fig 12: Lower side of the DDS PCB.

Construction

The component layout of the DDS PCB is shown in Fig 10, it is 55mm x 111mm and fits into a standard tinplate enclosure. Fig 11 shows the SMD component side of the PCB and Fig 12 shows the lower side of the PCB. The prototype PCB was commercially manufactured and is shown in the photo Fig 13. The crystal oscillator can be seen at the bottom left followed by the noise suppression circuit needed for the oscillator. The two-pole helix filter is above the oscillator followed by the buffer amplifier and the second helix filter.

The square black block is the DDS with the broadband transformer below and then the low pass filter with coupling capacitor to the output stage. The PCB was designed for 1812 size SMD components.

The most difficult part when assembling the PCB is soldering the DDS IC. It is not only the 48 fine spaced legs that have to be soldered but also the earth pad under the IC. Some amateurs drill an additional 3mm hole in the PCB under the IC so that it can be filled up with solder. The author uses for it the following procedure:

Coat the Pads of the printed circuit board with SMD paste, clamp the PCB on a table or in a vice, place the IC and aligning the legs to the pads of the PCB, heat the printed circuit board up from underneath with a hot-air blower until the tin solder contained in the paste melts. After the PCB has cooled surplus solder that bridges connections can be removed using de-soldering braid.

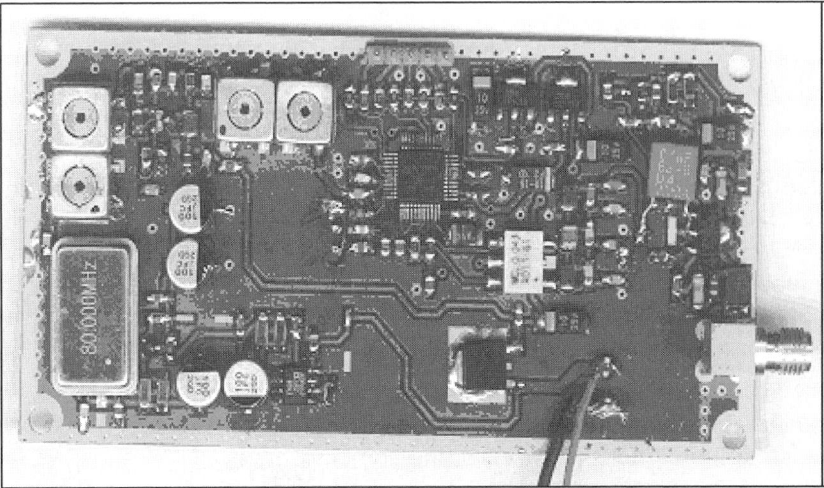

Fig 13: Photograph of the prototype DDS PCB.

Controlling this procedure is made much easier with a magnifying glass. Following this, fitting the remaining SMD components is child's play.

Measurements

The output spectrum and phase noise was measured up to 160MHz in the authors workshop starting from 10MHz, the lowest frequency usable on the E5052 analysers from Agilent.

Spectrum measurements

15 output spectrum plots of the DDS in 10MHz steps from 11 - 151MHz and 159MHz were made, too many to print in this book, they are available on the VHF Communivations magazine web site. The spurious level is remarkable it is only suppressed by 42dBc up to 100MHz (¼ the clock frequency). The harmonics of the desired signal are suppressed by at least 60dB.

Despite the low pass filters, leakage of the alias frequency can be seen when changing the frequency produced by the DDS. The alias frequency is about 100MHz away from the desired frequency. At an output frequency around 50MHz then signal and the alias frequency coincide. If the DDS is used as local oscillator for a short wave receiver with a high side intermediate frequency of 45MHz a birdie will appear on received signals at 5MHz that cannot be removed.

According to the data sheet the spurious level should be suppressed by better than 70dBc. The reason for problem was examined; the transformer on the output of the DDS did not seem to be balanced as shown when it was replaced. The difference in the phase noise on the fundamental of the crystal oscillator and the fifth harmonic is 14dB and corresponds to the theory.

In a discussion at the end of the lecture given by the author during the UKW conference at Bensheim 2007 [16], a listener reported that he had noticed the spurious level problem when developing a commercial product. The reasons were insufficient decoupling and insufficient bonding of the thermal pad. Better soldering of the thermal pad gives better spurious levels because the DAC in the DDS chip is bonded to the thermal pad. Attempts by the author with 100nF and 1nF decoupling capacitors on the supply line made no difference. More control by the author when soldering the thermal pad is not possible. Therefore the middle plated through hole under the thermal pad PCB shown here is 2mm diameter. After soldering the legs of the DDS this large hole can be filled with solder to ensure a good earth connection.

Phase noise measurements

15 phase noise plots of the DDS in 10MHz steps from 11 - 151MHz and 159MHz were made, too many to print in this book, they are available on the VHF Communications magazine web site.

At low output frequencies (e.g. 11MHz) the DDS phase noise is approximately 20dB lower than those of the 400MHz clock. At first sight this result is amazing because the DDS can be regarded like a frequency divider to explain the phase noise improvement. The relationship of 400MHz to 11MHz the improvement should be 20 log (400/11) = 31.2dB however only approximately 20dB was measured. The difference is caused because the frequency divider is not noise free. At an output frequency of 131MHz the phase noise curve is not good. The phase noise curve also flattens out for higher carrier frequencies.

This means that the DDS is not suitable as a local oscillator for a large signal 2m receiver because the phase noise is too high even at some distance from the carrier. Because the DDS has better phase noise close to the carrier than a VCO, the combination of a DDS and a VCO

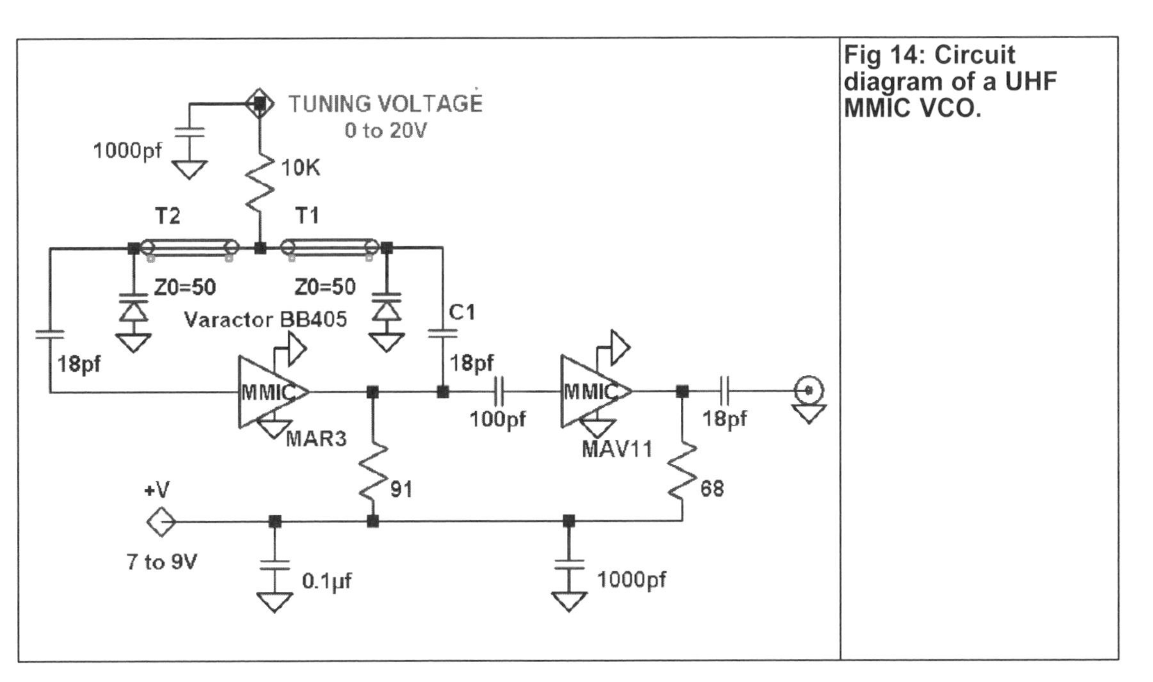

Fig 14: Circuit diagram of a UHF MMIC VCO.

may be considered. This is being investigated.

Conclusion

Since at the end of 2007 more DDS ICs in the same family with higher clock frequencies (1GHz) have become available from Analog Devices e.g. AD9910, AD9912. Signals up to 400MHz can be generated that can be used as an oscillator in 2m a transceiver. First publication appeared in Funkamateur Magazine also I0CG and WB6DHW have worked on a version. Because of the word length of the control data programming with an 8 bit controller is not easy.

MMIC Oscillator Experiments [19]

MMICs (Monolithic Microwave Integrated Circuits) are incredibly useful as amplifiers, but oscillate much less frequently than most other amplifiers. They are well-behaved enough to oscillate when desired. And they are predictable enough that I was able to build a wide-range VCO (Voltage Controlled Oscillator) that behaves nearly as simulated in software.

I have a fair amount of test equipment, but no signal generators that cover frequencies between 1300 and 2000MHz, at least ones that I can lift. So, when I wanted to test some 1296MHz filters, I had to come up with something. In the past, I have used a Minicircuits Frequency Doubler to cover this range, and I thought it might be easier if I could sweep the frequency with a VCO.

Some time ago, I played with a UHF VCO design from VHF Communications magazine (I've lost the reference - probably [20], but wasn't happy with the amplitude variation with frequency. At the time, it was a good lesson in why you shouldn't use a wide-range oscillator for a synthesised oscillator – small changes in tuning voltage cause large changes in frequency, so normal random noise results in a lot of phase noise. However, I recalled the technique used to vary the frequency and decided to try it with a MMIC.

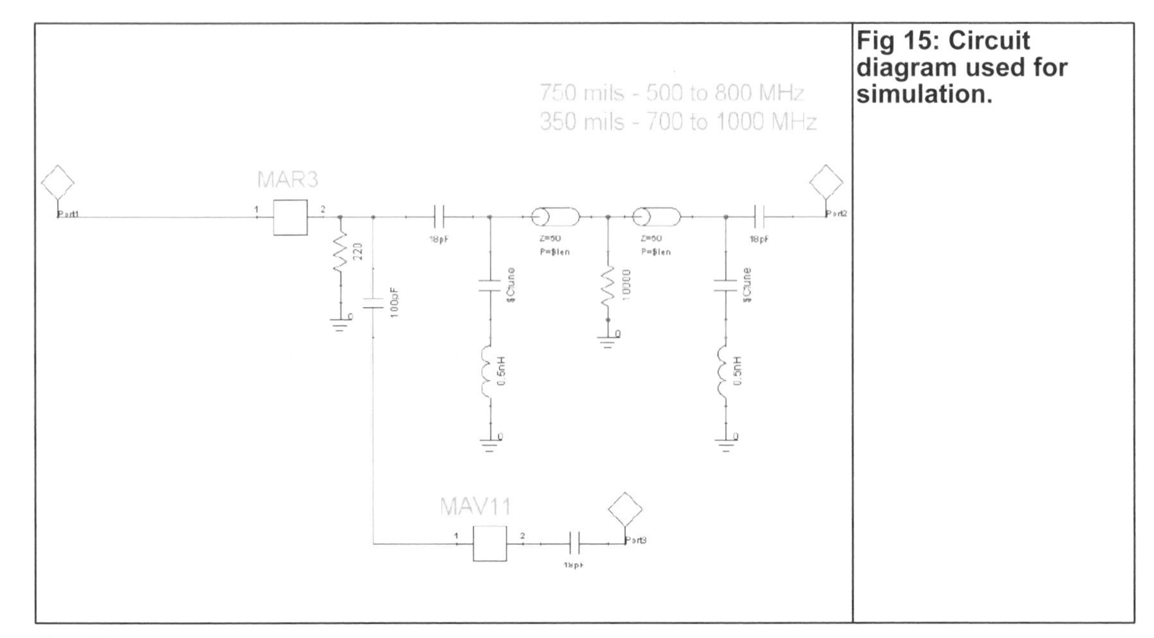

Fig 15: Circuit diagram used for simulation.

Oscillator

One type of oscillator uses positive feedback – we apply feedback around an active device to form a loop. The Barkhausen criteria states that when the loop gain is greater than one and the phase shift around the loop is 360 degrees, it will oscillate. An inverting device like a transistor or MMIC has 180 degrees phase shift at low frequencies, so we must provide the other 180 degrees – for instance, a half-wave line. At higher frequencies, the time an electron takes to get through a transistor becomes significant, so the internal phase shift is greater than 180 degrees and less is required externally. Conceptually, it is easy to make a single frequency oscillator with a half-wavelength of coax.

MMIC VCO

Making the frequency variable and controllable is where the difficulty lies. We would like to vary the length of the line forming the feedback loop, to control the phase shift. Adding capacitance to a transmission line makes it electrically longer. In the circuit in Fig 14, a varactor diode at each end of the transmission line acts as a variable capacitor, so that the line length is tunable. Increasing the tuning voltage reduces the capacitance, increasing the frequency. The tuning voltage is connected in the centre of the line, where it will have the least effect.

The buffer amplifier allows the oscillator to work into a relatively constant load, so that the frequency is less affected by whatever is attached to the output.

Simulation

The circuit was simulated using the free Ansoft Designer SV software [21]. Since our primary concern is loop gain and phase shift, we break the loop and look at it as an amplifier, as shown in Fig 15. The loop is the gain from Port1 to Port 2; we plot both gain and phase, looking at the gain where the phase shift is 360 degrees. The varactor tuning is approximated by changing the value of $Ctune – for the BB405 diode, the range is 2 to 11pf. With ¾ inches of transmission line, it should tune from about 500 to 800MHz. The simulation files will be available at [22].

One key to good VCO operation is that there is only one frequency with gain greater than 1 and 360 degrees phase shift. The MAR3 MMIC is a good choice for this frequency range – the gain drops off enough at the third harmonic so that loop gain is less than one, so the oscillator will only operate at one frequency. A hottter microwave device would have enough gain at higher harmonics to potentially oscillate at the third harmonic, where the transmission line is 3/2 wavelengths. While tuning, the oscillator might jump between fundamental and harmonic frequencies.

Test Results

I kludged a couple of these oscillators together on scraps of PC board cut from transverter prototypes (too ugly for photos!), using 0.141" semi-rigid coax for the transmission line.

With about ¾ inch of transmission line, the VCO tuned from about 375 to 650MHz, with about +19 dBm out of the buffer amplifier. A shorter transmission line, about ½ inch, tuned from about 380 to 700MHz, with about +16dBm output.

The output level is plenty to drive a surplus broadband doubler module – after a filter, an output power of 0 to +2dBm was available at twice the frequency.

The frequency range is somewhat lower than simulated, probably because the capacitance added by the PC board raises the minimum tuning capacitance. Dead-bug construction should have less stray capacitance and might work better. For lower frequency operation, longer lines and varactors with higher capacitance will do the job.

Microwave Oscillators

The same techniques could be used to make a loop oscillator for higher frequencies, by replacing the feedback loop with a pipe-cap filter. At resonance, phase changes very quickly, so there will be some frequency very close to resonance with the right phase shift and low enough loss to permit oscillation.

The MMIC should be placed between the two pipe-cap probes so that lead length is minimised. No blocking capacitors are needed since the probes provide DC isolation.

For a given probe length, filter coupling is reduced at lower frequencies. However, MMIC gain is higher at lower frequencies. Thus, it should be possible to find a probe length that works over the wide tuning range of the pipe-cap. My initial attempt was not quite right – it oscillates around 4GHz but not at lower frequencies.

Summary

MMIC oscillators work without much difficulty, and are easy and cheap to build. For those who like to experiment, these should be fun to tinker with.

References

[1] DDS using the AD9951, Henning C. Weddig, DK5LV, VHF Communications Magazine, 2/2009 pp 104 - 117

[2] News of the network analyser, Editor of Funkamateur, Funkamateur 9/2005

[3] Beta test of the DK3WX of network analyser, Peter Zenker, QRP-QTC, Funkamateur H1 2006

[4] Kit network analyser FA-NWT01: Construction and start-up, Norbert Grauopner, DL1SNG, Günter Borchert, DF5FC, Funkamateur 10 /2006 and 11/2006

[5] DDS VFO for 2m Transceiver, Günter Zobel, DM2DSN, Funkamateur 11/2005 and 12/2005

[6] www.dl5mgd.de/dds/AD9951.htm

[7] www.darc.de/c/DDS.pdf (DL5MGD Design)

[8] http://www.radioamatore. it/i0cg/ad9951.html

[9] http://www.it.geocities.com/giulianioi0cg/ad9951.html (alternative Download address for I0CG)

[10] Pic-A-star, Peter Rohdes, G3XP

[11] A Technical Tutorial on Digitally Signal Synthesis, Analog Devices 1999

[12] Data sheet AD9951, Analog Devices

[13] Description of the AD9954 evaluation Board, Analog Devices

[14] Data sheet IRU1010-18, International Rectifier

[15] Low-noise multipliers, Peter Vogl, DL1RQ, Conference proceedings for amateur radio conference Munich 2004

[16] DDS with the AD9951, Dipl. Ing. Henning Weddig, DK5LV, conference proceedings 52. UKW conference Bensheim 2007

[17] Local oscillator - DDS - AD9910 SSB noise, Martein Bakker, PA3AKE, http://www.xs4all.nl/~martein/pa3ake/hmode/dds_ad9910_amnoise.html

[18] Data sheet BFG135 NXP

[19] MMIC Oscillator E£xperiments, Paul Wade, W1GHZ, w1ghz@arrl.net. From Scatterpoint June 2009

[20] Voltage Controlled Tuned Wideband Oscillators, Jochen Jirmann, DB1NV, VHF Communications Magazine 4/1986 pp 214 - 221

[21] Ansoft Designer SV software - www.ansoft.com

[22] Simulation files available from - www.w1ghz.org

Index

C

D

E

F

G

H

I

L

M

N